AF287800

Marc Technau

On Beatty sets and some generalisations thereof

Marc Technau

On Beatty sets and some generalisations thereof

Würzburg
University Press

Dissertation, Julius-Maximilians-Universität Würzburg
Fakultät für Mathematik und Informatik, 2018
Gutachter: Prof. Dr. Jörn Steuding, Prof. Dr. Stephan Baier

Impressum

Julius-Maximilians-Universität Würzburg
Würzburg University Press
Universitätsbibliothek Würzburg
Am Hubland
D-97074 Würzburg
www.wup.uni-wuerzburg.de

© 2018 Würzburg University Press
Print on Demand

Coverdesign: Julia Bauer

ISBN 978-3-95826-088-7 (print)
ISBN 978-3-95826-089-4 (online)
URN urn:nbn:de:bvb:20-opus-163303

Acknowledgments

I would like to express my deep appreciation to all those who have enabled me to complete this thesis. First and foremost, my advisor, Jörn Steuding, deserves my gratitude for the support he extended to my cousin Niclas and me during our studies and for helping the both of us to pursue mathematical research. I thank him for paving many of the roads explored in this thesis, and for being forgiving about the instances in which I failed to walk on some other roads he intended for me.

Furthermore, I should like to express my indebtedness to my other co-authors for the pleasure of working with them. Although our joint work has not made it into this thesis, its progression has alleviated the times of frustration when my other mathematical enquiries seemed to have hit a wall.

More thanks are due to Stephan Baier who answered some questions of mine and graciously accepted to serve as second appraiser to this thesis.

I thank Christina Ostwald for supporting me with her Russian skills, thus rendering some papers accessible to me, which would otherwise have remained undeciphered.

Moreover, I thank my many friends who have endured me over the years in the office and elsewhere.

Finally, noting that my time here in Würzburg came in tandem with some teaching load and the resulting bureaucracy, I am happy to have had the insightful help of many colleagues. Amongst these I am especially grateful to Jens Jordan, Robert Hartmann and Uta Reyes.

Würzburg, February 2018   *Marc Technau*

Abstract

For Beatty sets $\mathcal{B}(\alpha, \beta) = \{n\alpha + \beta : n \in \mathbb{N}\}$ with irrational $\alpha > 1$ and $\beta \in \mathbb{R}$, and p prime and coprime to z, we investigate the problem of detecting points $(m, \tilde{m})$ on the modular hyperbola

$$\mathcal{H}_{z,p} = \{(m, \tilde{m}) \in \mathbb{Z}^2 \cap [1, p)^2 : m\tilde{m} \equiv z \mod p\}$$

with $\max\{m, \tilde{m}\}$ as small as possible, i.e., we obtain non-trivial estimates for

$$\min\{\max\{m, \tilde{m}\} : (m, \tilde{m}) \in \mathcal{H}_{z,p}, \, m \in \mathcal{B}(\alpha, \beta)\}$$

for certain α (see Theorem 2.9). The proof rests on new estimates for incomplete Kloosterman sums along $\mathcal{B}(\alpha, \beta)$ which are in turn obtained on supplying a method due to Banks and Shparlinski with a new estimate for the periodic autocorrelation of the finite sequence

$$0, \, \mathrm{e}_p(y\overline{1}), \, \mathrm{e}_p(y\overline{2}), \, \ldots, \, \mathrm{e}_p(y\overline{p-1}), \quad \text{with } y \text{ indivisible by } p,$$

($\overline{m}$ denoting the unique integer $m' \in [1, p)$ with $mm' \equiv 1 \mod p$ and $\mathrm{e}_p(x) = \exp(2\pi i x/p)$; see Proposition 2.4), the latter being obtained from adapting an argument due to Kloosterman.

Furthermore, we investigate sets of the shape $\{\lfloor m\alpha_1 + n\alpha_2 + \beta \rfloor : m, n \in \mathbb{N}\}$. We show that they are always contained in some ordinary Beatty set $\mathcal{B}(\tilde{\alpha}, \tilde{\beta})$ where we give admissible choices for $\tilde{\alpha}$ and $\tilde{\beta}$. Their respective complement $\mathscr{C}$ in this ordinary Beatty set is shown to be finite and bounds for the supremum of $\mathscr{C}$ are provided (see the theorems in Section 3.2 for the precise formulation of these results). The proofs are based on basic distribution properties of the sequence of fractional parts $\{n\alpha_1^{-1}\alpha_2\}$, $n = 1, 2, \ldots$, when $\alpha_1^{-1}\alpha_2$ is irrational, and appeal to the finiteness of the Frobenius number associated with a suitably chosen instance of the Frobenius coin problem otherwise.

Lastly, we generalise the definition of Beatty sets to imaginary quadratic number fields in a natural fashion. Assuming the number field in question to have class number 1, we are able to show that these Beatty-type sets contain infinitely many prime elements provided that the parameter corresponding to α from above is not contained in the number field (see Theorem 5.5). When the number field is $\mathbb{Q}(i)$, then, using the Hurwitz continued fraction expansion, we obtain a number field analogue (Theorem 5.7) of a previous result of Steuding and the author, who gave a Beatty set analogue of Linnik's famous theorem on the least prime number in an arithmetic progression. These results are obtained from number field analogues of classical results about the distribution of $\{p\vartheta\}$, $p = 2, 3, 5, 7, 11, \ldots$, $\vartheta \in \mathbb{R} \setminus \mathbb{Q}$, which were worked out recently by Baier for $\mathbb{Q}(i)$ using Harman's sieve method. We generalise these arguments to imaginary quadratic number fields with class number 1 with our main result stated as Theorem 4.4.

Contents

List of Figures

Nomenclature

The *Vinogradov symbols* $\ll$ and $\gg$ have their usual meaning, and if the implied constant depends on some parameter ϵ (say), then we usually write $\ll_\epsilon$ to draw attention to this fact. If we say that some variable be *fixed*, then this usually entails that we drop such subscripts even if the implied constant depends on said variable. The same comments apply to the *Landau symbols* $o(\,\cdot\,)$ and $O(\,\cdot\,)$. By $\mathscr{A} \subset \mathscr{B}$ we mean that the set $\mathscr{A}$ is a strict subset of $\mathscr{B}$ and we use $\subseteq$ when we allow equality to occur.

The following constitutes a list of common notations used throughout this thesis:

General notation

$\mathbf{1}_{\{P\}}$	equals 1 if the proposition P holds true and equals 0 otherwise.
$\mathbf{1}_{\mathscr{X}}$	denotes the characteristic function of the set $\mathscr{X}$.
$\boldsymbol{\alpha}$	$= (\alpha_1, \alpha_2)$ where $\alpha_1, \alpha_2 \geq 1$ are real numbers.
$\arg \rho$	denotes the unique real number $t \in (-\pi, \pi]$ such that $\rho/|\rho| = \exp(it)$ if $\rho \neq 0$ and $\arg 0 = 0$.
$\mathcal{B}(\alpha, \beta)$	$= \{\lfloor n\alpha + \beta \rfloor : n \in \mathbb{N}\}$, the Beatty set associated with α and β.
$\mathcal{B}_\omega(\alpha, \beta)$	$= \{\lfloor n\alpha + \beta \rfloor_\omega : n \in \mathcal{O}\}$, the Beatty-type set assoc. w. α, β and ω.
$\mathcal{B}_{\mathrm{ts}}(\alpha, \beta)$	$= \{\lfloor n\alpha + \beta \rfloor : n \in \mathbb{Z}\}$, the two-sided Beatty set assoc. w. α and β.
$\operatorname{cosup}_{\mathscr{Y}} \mathscr{X}$	is the supremum of $\mathscr{Y} \setminus \mathscr{X}$.
$\operatorname{diam} \mathscr{R}$	$= \sup\{|x - y| : x, y \in \mathscr{R}\}$, the diameter of $\mathscr{R} \subseteq \mathbb{C}$.
$d_k(n)$	is the number of ways of writing n as a product of k positive integers; $d(n) = d_2(n)$.
$d \mid x$	means that d divides x.
$\mathrm{e}(x)$	$= \exp(2\pi i x)$.
$\mathrm{e}_p(x)$	$= \mathrm{e}(x/p) = \exp(2\pi i x/p)$ (with p prime).
$f(x) \asymp g(x)$	means $f(x) \ll g(x)$ and $g(x) \ll f(x)$.
$f(x) \sim g(x)$	means $f(x) = g(x)(1 + o(1))$ as $x \to \infty$.
G	$= \frac{1}{2}(\sqrt{5} + 1)$, the Golden ratio.

$\mathbb{N}, \mathbb{Z}, \mathbb{R}, \mathbb{C}$	are the sets of positive integers, integers, real numbers and complex numbers, respectively.
$\mathbb{N}_0$	$= \mathbb{N} \cup \{0\}$, the set of non-negative integers.
$O^*(\,\cdot\,)$	is the usual Landau $O(\,\cdot\,)$ with the additional requirement that the implied constant always be ≤ 1.
p	usually denotes a positive rational prime or a prime element in $\mathcal{O}$.
$\varphi(n)$	$= \#\{1 \leq m \leq n : m \text{ coprime to } n\}$.
$\pi(x)$	$= \#\{p \leq x : p \text{ prime}\}$.
$\Re_\omega x, \Im_\omega x$	are the unique real numbers such that $x = \Re_\omega x + (\Im_\omega x)\omega$.
$\Re x, \Im x$	denote the real part and imaginary part of x respectively.
$\mathrm{sgn}\, x$	is the sign of x, i.e., $\mathrm{sgn}\, x = +1$ if $x > 0$, $= -1$ if $x < 0$, and $= 0$ if $x = 0$.
$\#\mathcal{X}$	denotes the number of elements in the set $\mathcal{X}$.
$|x|$	is the absolute value of $x \in \mathbb{C}$.
$\{x\}$	$= x - \lfloor x \rfloor$, the fractional part of x.
$\lfloor x \rfloor$	is the largest integer less than or equal to x.
$\lfloor x \rfloor_\omega$	$= \lfloor \Re_\omega x \rfloor + \lfloor \Im_\omega x \rfloor \omega$.
$\|x\|$	$= \min_{y \in \mathbb{Z}}|x - y|$, the distance of $x \in \mathbb{R}$ to a nearest integer.
$\overline{x}$	may either be the unique integer y with $1 \leq y < p$ and $xy \equiv 1 \bmod p$ (used throughout Chapter 2) or the complex conjugate of x.

Number field related notation

$\mathfrak{a}$	usually denotes an ideal in $\mathcal{O}$.
$\mathscr{A}$	is usually some subset of $\mathscr{B}$ (in the context of Harman's sieve).
$\mathscr{B}$	$= \{m \in \mathcal{O} : x/2 \leq N(m) < x\}$ (in the context of Harman's sieve).
$d_k(\mathfrak{a})$	is the number of ways of writing the ideal $\mathfrak{a}$ as a product of k ideals; $d(\mathfrak{a}) = d_2(\mathfrak{a})$.
$\mathbb{K}$	is an imaginary quadratic number field embedded into $\mathbb{C}$.
$M_\mathcal{O}(\epsilon)$	is the number given in Lemma A.8.

(n)	is the principal ideal in $\mathcal{O}$ generated by $n \in \mathcal{O}$.
$N(m)$	$= \#(\mathcal{O}/m\mathcal{O}) = |m|^2$, is the the norm of $m \in \mathcal{O}$; if $\mathfrak{a} \subseteq \mathcal{O}$ is an ideal, then we just write $N\mathfrak{a}$ for the number of elements in $\mathcal{O}/\mathfrak{a}$.
$\mathcal{O}$	is the ring of integers of $\mathbb{K}$.
ω	$= \Omega_1 + \Omega_2 i$ usually is some fixed generator of $\mathcal{O}$.
$\mathfrak{p}$	usually denotes a prime ideal in $\mathcal{O}$.
$\|x\|_\omega$	$= \max\{\|\Re_\omega x\|, \|\Im_\omega x\|\}$.
ξ_1, ξ_2	are given by $\Re_\omega \omega^2$ and $\Im_\omega \omega^2$ respectively; $\omega^2 = \xi_1 + \xi_2\omega$.

Summation conventions

$\sum_{n \leq x}$	sum over all positive integers $n \leq x$.
$\sum_{0 \leq n \leq x}$	sum over all non-negative integers $n \leq x$.
$\sum_{0 \leq |j| \leq J}$	sum over all integers j with $|j| \leq J$.
$\sum_{d|x}$	sum over all divisors d of x; if $a, b \mid x$ are associate, then d only assumes the value of either a or b.—The terms being summed will not ever depend on the particular choice of associate.

Additional conventions

Moreover, we have tried within reason to stick to some general notational guidelines and hope that their explicit exclamation here will increase the readability rather than startle the reader at those points where we have failed to abide by them:

- We often use caligraphic letters $\mathscr{A}, \mathscr{B}, \mathscr{C}, \ldots$ to denote sets.

- In the context of imaginary quadratic number fields $\mathbb{K} = \mathbb{Q}(\omega)$ with ring of integers $\mathcal{O}$, given some element $\vartheta \in \mathbb{K}$, we often write ϑ_1 for $\Re_\omega \vartheta$ and ϑ_2 for $\Im_\omega \vartheta$. In particular, if we state that (say) ϑ_1 and ϑ_2 be such that $\vartheta = \vartheta_1 + \vartheta_2\omega$, it is *always* implicit that they be real.

- In view of our bias towards $\vartheta_1 = \Re_\omega \vartheta$, when in need of two ϑs, we usually favour adding a tilde, thus writing ϑ and $\tilde{\vartheta}$ instead of, for instance, ϑ_1 and ϑ_2.

Chapter 1

Introduction

The principal object of study in this thesis are certain properties of so-called Beatty sequences and two proposed variants thereof to be defined below. However, before shedding light on the specifics (see Section 1.2), we attempt to paint a small panorama of facets of Beatty sequences that have been explored in the literature; certainly we make no attempt at an exhaustive account of the available material, but perhaps our selection will prove sufficient for sparking the reader's interest in what is to be found in subsequent chapters. The style in Section 1.1 is quite leisurely.

1.1 Facets of Beatty sequences

1.1.1 The Rayleigh–Beatty theorem

We set the stage with the following problem proposed for solution in 1926 by Samuel Beatty in *The American Mathematical Monthly* [16, 17]:

> "If X is a positive irrational number and Y its reciprocal, prove that the sequences
>
> $$(1+X), \quad 2(1+X), \quad 3(1+X), \quad \ldots$$
> $$(1+Y), \quad 2(1+Y), \quad 3(1+Y), \quad \ldots \tag{1.1}$$
>
> contain one and only one number between each pair of consecutive positive integers."

Although the publication of the above result—often referred to as *Beatty's theorem*—in the *Monthly* popularised the sequences given in (1.1), the result itself actually seems to be due to John William Strutt (3rd Baron Rayleigh) who provided the following equivalent formulation of it in the 2nd edition of volume 1 of his treatise on *The Theory of Sound* from 1894 (see [80, §92a]):

> "[I]f x be an incommensurable number less than unity, one of the series of quantities m/x, $m/(1-x)$, where m is a whole number, can be found which shall lie between any given consecutive integers, and but one such quantity can be found."

In view of the above, it has become customary in the literature to consider sets of the form

$$\mathcal{B}(\alpha, \beta) = \{\lfloor n\alpha + \beta \rfloor : n \in \mathbb{N}\}, \tag{1.2}$$

1

where α (the *slope*) and β (the *shift*) are real numbers and $\lfloor x \rfloor$ denotes the largest integer less than or equal to x, and we note in passing that such sets appeared already in [20] (see also [62]). The set $\mathcal{B}(\alpha, \beta)$ is called *Beatty set* associated with α and β or *Beatty sequence* where it is understood that the elements of $\mathcal{B}(\alpha, \beta)$ are to be enumerated in increasing order. Often enough only the case $\beta = 0$ is considered and, thus, to stress the presence of the parameter β, the sets $\mathcal{B}(\alpha, \beta)$ are sometimes referred to as *inhomogeneous* Beatty sets, contrasting them with the *homogeneous* Beatty sets $\mathcal{B}(\alpha, 0)$.[1]

In accordance with the usual number theorist's bias towards favouring positive integers over the set of all integers, some authors require $\alpha \geq 1$ and $\beta \geq 0$. However, disregarding these restrictions and even dropping this bias in (1.2), thus considering the *two-sided* Beatty sets (or *Beatty bisequences*)

$$\mathcal{B}_{\text{ts}}(\alpha, \beta) = \{\lfloor n\alpha + \beta \rfloor : n \in \mathbb{Z}\}, \tag{1.3}$$

also leads to interesting results (see, e.g., [73])—an attitude to which we shall return shortly when defining Beatty-type sets in imaginary quadratic number fields below.

Returning to our initial discussion of the Rayleigh–Beatty theorem, we note that several generalisations have been suggested, among those we wish to highlight the work of Fraenkel [25] who studied the inhomogeneous case, and Lambek and Moser [55] who obtained a general criterion, for checking if two (potentially finite) sequences partition $\mathbb{N}$, which, amongst other things, implies the Rayleigh–Beatty theorem. Moreover, Skolem [75] (see also [7]) rediscovered the Rayleigh–Beatty theorem and proved that there cannot be more than two mutually disjoint Beatty sets.

1.1.2 Intersecting Beatty sets and gap theorems

We mention another surprising phenomenon connected with Beatty sets: for instance, one may check that the intersection, $\mathscr{X}$ say, of $\mathcal{B}(\sqrt{3}, 0)$ with the arithmetic progression $\mathcal{B}(3, 0)$ contains infinitely many elements. Clearly the sequence obtained from enumerating the elements of $\mathscr{X}$ in increasing order cannot be periodic. Thus it may come as a surprise that the *sizes of gaps* in that sequence (the lengths of the intervals bounded by consecutive elements of the sequence) is very regular: only gaps of the sizes 3, 6 and $9 = 3 + 6$ occur (see Fig. 1).

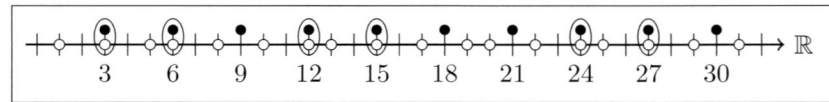

Figure 1: Intersecting $\mathcal{B}(\sqrt{3}, 0)$ (points on the line) with the arithmetic progression $\mathcal{B}(3, 0)$ (points above the line).

This is a special case of the observation of Fraenkel and Holzman [26], who noted that upon intersecting Beatty sets with arithmetic progressions at most three distinct

---

[1] In particular, the reader should be aware that some of the sources cited in this section use more restrictive definitions of what a Beatty sequence is supposed to be.

gap sizes occur and, moreover, if three do occur, then one of the gap sizes is the sum of the other two. They compare this to two problems:

- First, the so-called *Steinhaus problem* which asks for the maximum number of distinct sector sizes that can appear when partitioning the unit circle into sectors bounded by the points

$$\exp(2\pi in\alpha), \quad n = 1, 2, \ldots, N,$$

  where α is real and N a positive integer (see Fig. 2). (Hugo Steinhaus conjectured the answer to this to be three and this was verified independently by Sós [76] and Świerczkowski [81].—The result became known as the *Three Gap Theorem*.)

- Second, the so-called *Slater problem* which asks the following: let $\mathscr{I}$ be any interval modulo one and let $n_1 < n_2 < \ldots$ be all those (possibly finite) positive integers n for which the *fractional part* $\{n\alpha + \beta\}$ belongs to $\mathscr{I}$. How many distinct gap lengths

$$n_{k+1} - n_k, \quad k = 1, 2, \ldots$$

  do occur? (Again, the answer turns out to be "at most three" and we refer to [26] and the references therein.)

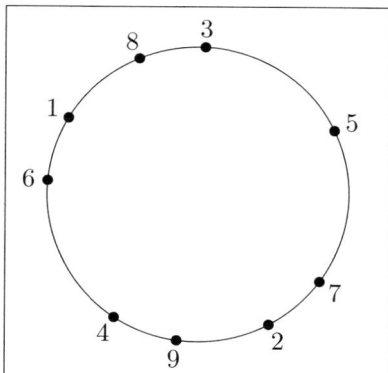 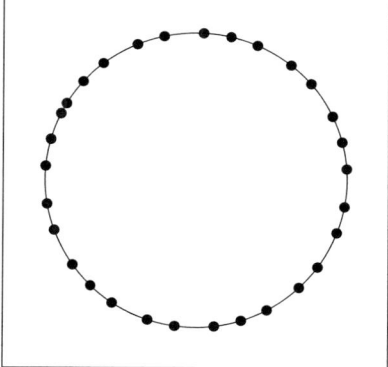

Figure 2: An illustration of the Three Gap Theorem with $\alpha = \sqrt{2}$ and $N = 9$ (left) and $N = 30$ (right). The circle on the left also shows the numbers n next to the points $\exp(2\pi in\alpha)$.

For more on this, we refer to [26] (see also [27]).—Suffice it to say that they also study the more general situation of intersecting arbitrary Beatty sets and also obtain finiteness results for the number of possible gaps, although the number three has to be adjusted.

1.1.3 Arithmetic functions on Beatty sets

It is certainly a very well established idea (in analytic number theory in particular) that when searching for certain elements, of arithmetic interest (say), in a subset of the positive integers, the failure of being able to exhibiting such elements explicitly can be remedied by looking at averages. Therefore, given some arithmetic function $f : \mathbb{N} \to \mathbb{C}$ of interest, one might like to understand the behaviour of

$$\sum_{\substack{m \leq x \\ m \in \mathcal{B}(\alpha,\beta)}} f(m) \quad \text{as } x \to \infty. \tag{1.4}$$

Since $\mathcal{B}(\alpha, \beta)$ turns out to be the union of arithmetic progressions if α is rational, such considerations often fall into classical territory of analytic number theory and are well-studied within that context. Hence, the primary interest here lies with irrational α and the quality of available estimates generally depends in one way or another on properties of Diophantine approximability of α. Subject to these remarks, there are results for a multitude of particular choices for the function f and irrational α:

- d, the *divisor function* (i.e., the function giving the number of ways of writing its argument as a product of two positive integers), starting with Abercrombie [1], improved by Begunts [18], and work on generalised divisor functions by Zhai [93] and Lü and Zhai [61].

- Certain multiplicative functions like $n \mapsto n^{-1}\varphi(n)$, where φ is *Euler's totient*, due to Begunts [19], and a certain class of multiplicative f in the work of Güloğlu and Nevans [30].

- Dirichlet-characters and special functions related to the orbits of elements $g \bmod m$ along Beatty sequences, treated by Banks and Shparlinski [10] (mind also [11]), with improvements when f is the Legendre-Symbol due to Banks, Garaev, Heath-Brown, and Shparlinski [8].

- ω, the function counting the number of distinct prime divisors of its argument (without multiplicity), and $n \mapsto (-1)^{\Omega(n)}$, where $\Omega(n)$ counts the number of distinct prime divisors of n with multiplicity, due to Banks and Shparlinski [12].

- Λ, the *von Mangoldt lambda function*, studied by Banks and Shparlinski [13] who attribute earlier results to Ribenboim [69, Chapter 4.V], although it seems that such observations were already apparent to Heilbronn in 1954, as evidenced in [89, Notes to Chapter XI].

The general heuristic is that—unless one has some strong reason to believe otherwise—the arguments $m \in \mathcal{B}(\alpha, \beta)$, $m \leq x$, should sample the function f in an unbiased fashion. In particular, fixing α and β, one should generally expect that (1.4) be asymptotic to

$$\frac{1}{\alpha} \sum_{m \leq x} f(m), \tag{1.5}$$

where the factor α^{-1} accounts for the fact that $\alpha^{-1}x(1 + o(1))$ integers $n \leq x$ belong to $\mathcal{B}(\alpha, \beta)$.

One may contrast this, for instance, with the *Titchmarsh divisor problem* raised by Titchmarsh [82, 83], which asks for the behaviour of

$$\mathrm{Dp}(x; \ell) = \sum_{\ell < p \leq x} d(p - \ell) \quad \text{as } x \to \infty,$$

where the variable p only assumes prime values. Titchmarsh provided upper and lower bounds for this quantity (unconditionally) and, on assuming the extended Riemann hypothesis, showed that

$$\mathrm{Dp}(x; \ell) \sim \delta(\ell) x \quad \text{as } x \to \infty,$$

where

$$\delta(\ell) = \prod_{(p,\ell)=1} \left(1 + \frac{1}{p(p-1)}\right) \prod_{q \mid \ell} \left(1 - \frac{1}{q}\right) \quad \text{(both } p \text{ and } q \text{ prime).}$$

Linnik [59] later obtained unconditionally

$$\mathrm{Dp}(x; 1) \sim \frac{315\zeta(3)}{2\pi^4} x,$$

where ζ denotes the Riemann zeta function. (See also [72, § II.11] for a collection of more recent work on the problem.) In any case, recalling the well-known formula for the average number of divisors of the intergers $\leq x$, one has

$$\frac{1}{\pi(x)} \sum_{p \leq x} d(p - 1) \not\sim \frac{1}{x} \sum_{m \leq x} d(m),$$

in violation of what one may come to expect by naively appealing to the analogue of the aforementioned heuristic. (Here $\pi(x)$ denotes the number of primes $p \leq x$.)

Clearly, also the above heuristic presented for Beatty sequences may fail for particular choices of α, β and f (take, e.g., f to be $\mathbf{1}_{\mathbb{Z}\setminus\mathcal{B}(\alpha,\beta)}$, the characteristic function of $\mathbb{Z} \setminus \mathcal{B}(\alpha, \beta)$), but Abercrombie, Banks, and Shparlinski [2] show that the heuristic works in a metric sense provided that f does not grow too quickly. We give the full statement of their result, as we shall refer to it later:

Theorem 1.1 (Abercrombie, Banks, and Shparlinski [2]). *Let f be some arithmetic function and $\epsilon > 0$. Then, for almost all real numbers $\alpha > 1$ (in the sense of Lebesgue measure),*

$$\left| \sum_{\substack{n \leq x \\ n \in \mathcal{B}(\alpha,0)}} f(n) - \frac{1}{\alpha} \sum_{n \leq x} f(n) \right| \ll_{f,\alpha,\epsilon} x^{2/3+\epsilon} \left(1 + \max_{n \leq x} |f(n)|\right).$$

Of course, strictly speaking one must also require the average of f not to be too small if one intends to derive from this, that (1.4) be asymptotically equal to (1.5). However, even if such a result cannot be obtained along these lines, settling for some non-trivial bound for (1.4) obtained via Theorem 1.1 may suffice for applications.

5

1.1.4 Primes and Beatty sets

A line of research concerned with the interrelation between primes and Beatty sets was already foreshadowed in Section 1.1.3 by our mention of the von Mangoldt lambda function.

For instance, one might ask to characterise the values of $\alpha \geq 1$ and $\beta \geq 0$ for which the set $\mathcal{B}(\alpha, \beta)$ contains infinitely many primes. When α is an integer, appealing to Dirichlet's celebrated theorem on primes in arithmetic progressions, this reduces to asking if α and $\lfloor \beta \rfloor$ are coprime. When $\alpha = \frac{a}{q}$ is a reduced fraction with $2 \leq q < a$, the answer looks similar: one easily observes that $\mathcal{B}(\alpha, \beta)$ splits into a disjoint union of arithmetic progressions,

$$\mathcal{B}(\tfrac{a}{q}, \beta) = \bigcup_{1 \leq b \leq q} \left(\left\lfloor \tfrac{ab}{q} + \beta \right\rfloor + \mathbb{N}_0 a \right). \tag{1.6}$$

Thus, the question about infinitely many primes in $\mathcal{B}(\alpha, \beta)$ reduces to asking whether there is some b such that $\lfloor \frac{ab}{q} + \beta \rfloor$ and a are coprime (for instance, Ribenboim [69, Chapter 4.V] notes that for $\beta = 0$ there is always such a b; indeed, in that case, taking b to be the multiplicative inverse of a modulo q does the job, as remarked in [79]). Certainly it would be desirable to have a more direct characterisation of those β allowing for the existence of some b with the aforementioned property.

Anyway, moving on to the case of irrational α, it turns out that $\mathcal{B}(\alpha, \beta)$ *always* contains infinitely many primes. To see this, one may use the following simple characterisation of membership to the Beatty set, which can be proved quite directly from just the definition of $\mathcal{B}(\alpha, \beta)$:

Lemma 1.2. *Suppose that $\alpha \geq 1$ and $\beta \geq 0$ are real numbers. Then, for any integer m,*

$$m \in \mathcal{B}(\alpha, \beta) \quad \textit{if and only if} \quad \begin{cases} \dfrac{m}{\alpha} \in \left(\dfrac{\beta - 1}{\alpha}, \dfrac{\beta}{\alpha} \right] \bmod 1, \\ m > \alpha + \beta - 1. \end{cases}$$

(Here, given some subset $\mathscr{F}$ of the real numbers, $\mathscr{F} \bmod 1$ denotes the union of all $\mathbb{Z}$-translates of $\mathscr{F}$.)

From works of I. M. Vinogradov (see [89, Chapter XI] for an exposition thereof and also [88]) and known theorems on the distribution of primes in arithmetic progressions one readily deduces the following theorem:

Theorem 1.3. *Let p_n denote the n-th prime and suppose that α is irrational. Then the sequence $(\{p_n/\alpha\})_{n \in \mathbb{N}}$ is uniformly distributed modulo one.*

For an account on the theory of uniform distribution we refer the reader to [53] and for more details on the proof of the above result see [45, p. 489]. Using the above, one immediately obtains

$$\lim_{x \to \infty} \frac{1}{\pi(x)} \# \left\{ p \leq x \text{ prime} : \frac{p}{\alpha} \in \left(\frac{\beta - 1}{\alpha}, \frac{\beta}{\alpha} \right] \bmod 1 \right\} = \frac{1}{\alpha}, \tag{1.7}$$

that is, a prime number theorem for $\mathcal{B}(\alpha, \beta)$. For $\beta = 0$ this line of reasoning was already given at least as early as in 1954 by Heilbronn (again, see [89, Notes to Chapter XI]), although we do not wish to preclude the possibility of this being common knowledge at that time and, perhaps, not having originated with Heilbronn. (In any case, we are just talking about a straightforward rephrasing of some of Vinogradov's results here, so the correct attribution of the connection with Beatty sets is probably only of minor interest to begin with.) We also refer to [36] for a discussion of primes in intersections of Beatty sequences.

Now given the fact that there are infinitely many primes in $\mathcal{B}(\alpha, \beta)$ when α is irrational—and even quite many at that (see (1.7))—it is natural to investigate which classical questions about primes in $\mathbb{N}$ or arithmetic progressions can also be settled for Beatty sequences. Along those lines we remark that, amongst other things, the following topics have been investigated with respect to their Beatty-analogues:

- Goldbach's problem [14, 54, 60, 87] and variants [64],

- Primes in short intervals [24, 37],

- Waring's problem [15],

- Gaps between primes [6] (see also [9]),

- Linnik's theorem [79].

We shall dwell on the last point some more, because it serves as a motivation for some of the later work in this thesis. More specifically, by *Linnik's theorem* we mean the following result, considered by some to be *"one of the greatest achievements in analytic number theory"* [45, p. 427]:

Theorem 1.4 (Linnik [57, 58]). *There is some constant $L \geq 2$ such that for any coprime integers a, b ($a \geq 2$) the least prime number $p \equiv b \bmod a$ satisfies the bound $p \ll_L a^L$, where the implied constant depends on L alone.*

Although Linnik himself did not bother to explicitly compute an admissible value for L, others have since spent quite a lot of thought on that topic; the current record seems to be $L = 5$ due to Xylouris [92] who drew significantly on the earlier work of Heath-Brown [38], who obtained $L = 5.5$ (see also [38, Table 1] for a list of prior work on the admissible value of L).

Returning to Beatty sequences, we note that, by picking some $\alpha > 1$ with composite integer part $\lfloor \alpha \rfloor$ one immediately observes that the associated Beatty sequence $\mathcal{B}(\alpha, \beta)$ can be made to start with arbitrarily many composite numbers, provided that the fractional part $\{\alpha\}$ be small enough. Therefore, one should not be looking to bound the smallest prime in $\mathcal{B}(\alpha, \beta)$ in terms (say) $\lfloor \alpha \rfloor$ alone. However, a straightforward argument (which we shall sketch below for the sake of completeness), gives the following result:

Proposition 1.5. *Suppose that L is as in Theorem 1.4. Then, if the Beatty sequence $\mathcal{B}(\frac{a}{q}, \beta)$ ($a > q > 1$ coprime, $\beta \geq 0$) contains a prime at all, the least such prime p satisfies the bound $p \ll_L (1 + \beta)^L a^{2L}$, where the implied constant only depends on L.*

Clearly this is in the spirit of Linnik's theorem! Concerning irrational slope α, Steuding and the author [79] proved the following:

Theorem 1.6 (Steuding and T. [79]). *For every $\epsilon > 0$ there exists a positive integer ℓ such that, for every irrational $\alpha > 1$ and real $\beta \geq 0$, the least prime p in the Beatty sequence $\mathcal{B}(\alpha, \beta)$ satisfies the inequality*

$$p \leq \mathcal{L}^{35-16\epsilon} \alpha^{2(1-\epsilon)} B a_{m+\ell}^{1+\epsilon},$$

where $B = \max\{1, \beta\}$, $\mathcal{L} = \log(2\alpha B)$, a_n denotes the numerator of the n-th convergent to the regular continued fraction expansion of α and m is the unique integer such that $a_m \leq \mathcal{L}^{16}\alpha^2 < a_{m+1}$.

Proof of Proposition 1.5. Consider (1.6). If there is no b ($1 \leq b \leq q$) such that $x(b) = \lfloor \frac{ab}{q} + \beta \rfloor$ is coprime to a, then all primes in $\mathcal{B}(\frac{a}{q}, \beta)$, if any, are of the form $x(b) \leq a + \beta < (1+\beta)a$ with $1 \leq b \leq q$. Hence, the assertion of the proposition holds in this case.

Now suppose that there is some b ($1 \leq b \leq q$) such that $x(b)$ is coprime to a. Clearly, by Linnik's theorem, there is some prime $\ll a^L$ in the residue class $x \bmod a$, but this prime may fail to belong to the Beatty set due to the fact that the arithmetic progressions on the right hand side of (1.6) do not necessarily contain all positive integers in the corresponding residue class. To get around this problem, we put $y = 1 + a(\lfloor \beta \rfloor + 1)$ and employ the Chinese remainder theorem to find some integer B coprime to $A = ay$ such that

$$\begin{cases} B \equiv x(b) \mod a, \\ B \equiv 1 \mod y. \end{cases} \tag{1.8}$$

Linnik's theorem then ensures the existence of a prime $p \equiv B \bmod A$ with

$$p \ll A^L \ll_L (1+\beta)^L a^{2L}.$$

The second congruence in (1.8) ensures $p > y > a + \lfloor \beta \rfloor \geq x(b)$. Thus,

$$p \in (x(b) + \mathbb{N}_0 a) \subseteq \mathcal{B}(\tfrac{a}{q}, \beta). \qquad \square$$

1.2 Contribution of this thesis

The rest of this thesis essentially splits into three parts. We refer to Fig. 3 for a basic glimpse at the logical dependence amongst the chapters and end this chapter with a short explanation of what is to be encountered in the subsequent chapters.

1.2.1 A connection with modular inversion

In Chapter 2 we investigate the distribution of points on modular hyperbolas with one coordinate restricted to some Beatty set, that is, for p prime and z indivisible by p, we try to gain a better understanding of the distribution of the points $(m, \tilde{m})$ with

$$m\tilde{m} \equiv z \mod p, \quad m \in \mathcal{B}(\alpha, \beta) \quad \text{and} \quad 1 \leq m, \tilde{m} < p.$$

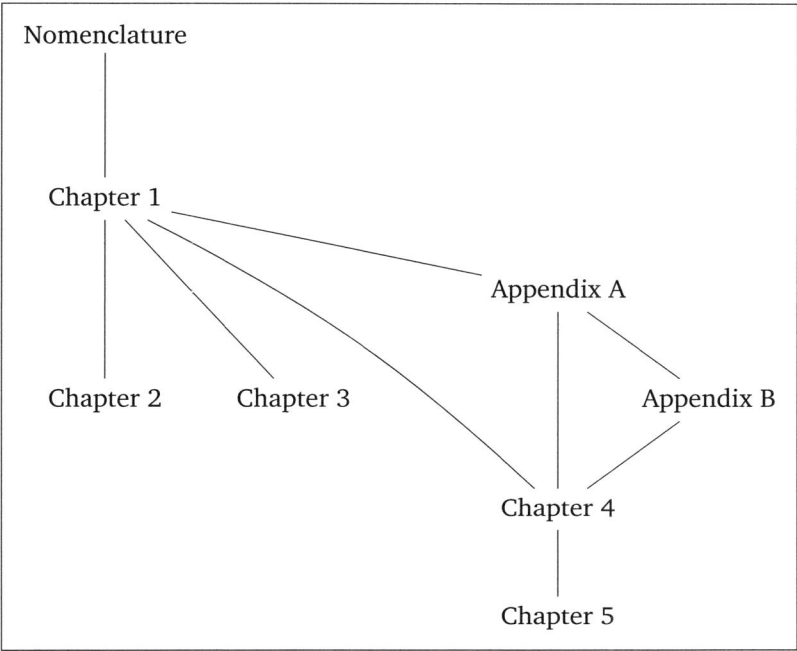

Figure 3: Logical dependence amongst the chapters.

The particular problem to be introduced in Section 2.1 of Chapter 2 turns out to be connected to bounding incomplete Kloosterman sums along points from $\mathcal{B}(\alpha, \beta)$. By adapting a method due to Banks and Shparlinski [10, 11, 12] and supplying additional arithmetic input (following Kloosterman [50]), we are able to bound such sums in certain ranges.

1.2.2 Denser Beatty sets

Chapter 3 originally grew out of the desire to construct a setting in which a stronger analogue of Theorem 1.6 could be obtained. The underlying line of thought was that enlarging the Beatty set $\mathcal{B}(\alpha_1, \alpha_2 + \beta)$ $(\alpha_1, \alpha_2 \geq 1, \beta \geq 0)$, that is, replacing it by the presumably "denser" set

$$\mathcal{B}(\boldsymbol{\alpha}, \beta) = \{\lfloor m\alpha_1 + n\alpha_2 + \beta \rfloor : m, n \in \mathbb{N}\} \tag{1.9}$$

(where $\boldsymbol{\alpha} = (\alpha_1, \alpha_2)$), should allow for some improved bounds on the least prime in such a set. As it turns out, however, sets of the type (1.9) actually equal ordinary Beatty sets save for finitely many exceptions and we contend ourselves with providing upper bounds for the largest such exception; as a consequence of this, the question about the least prime in such sets falls within the scope of Proposition 1.5 and Theorem 1.6 simply by applying them with a large enough shift β as to ensure that the "detected" prime cannot possibly be one the aforementioned exceptions. At the end of Chapter 3 we also discuss an interesting related problem concerning large gaps in the Frobenius coin problem.

1.2.3 Generalisation to imaginary quadratic number fields

All remaining chapters are quite closely related, the overall goal being to detect prime elements in Beatty-type sets in imaginary quadratic number fields; before defining these sets we note that in this more algebraic setting it seems more natural to drop the usual bias towards $\mathbb{N}$. We thus seek to generalise the two-sided Beatty sets (1.3). In this regard let us consider a full lattice $\mathcal{O} \subseteq \mathbb{C} \cong \mathbb{R}^2$ generated by two elements, $\mathcal{O} = \langle \tilde{\omega}, \omega \rangle$ say, and define an analogue of the usual floor function $\lfloor \cdot \rfloor : \mathbb{R} \longrightarrow \mathbb{Z}$ as

$$\lfloor \cdot \rfloor_{\tilde{\omega},\omega} : \mathbb{C} \longrightarrow \mathcal{O}, \quad x_1\tilde{\omega} + x_2\omega \longmapsto \lfloor x_1 \rfloor \tilde{\omega} + \lfloor x_2 \rfloor \omega \quad (x_1, x_2 \in \mathbb{R}).$$

Clearly, just as $\lfloor \cdot \rfloor$ is a retraction of the inclusion $\mathbb{Z} \longrightarrow \mathbb{R}$, $\lfloor \cdot \rfloor_{\tilde{\omega},\omega}$ is a retraction of the inclusion $\mathcal{O} \longrightarrow \mathbb{C}$. Moreover, for any $n \in \mathcal{O}$ and $\vartheta \in \mathbb{C}$,

$$\lfloor n + \vartheta \rfloor_{\tilde{\omega},\omega} = n + \lfloor \vartheta \rfloor_{\tilde{\omega},\omega}.$$

Now consider an imaginary quadratic number field $\mathbb{K}$ (which we always assume to be embedded in $\mathbb{C}$) and its ring of integers $\mathcal{O} = \mathbb{Z}[\omega]$. In view of $\dim_{\mathbb{R}}(\mathcal{O} \otimes_{\mathbb{Z}} \mathbb{R}) = 2$ one then obtains a function $\lfloor \cdot \rfloor_\omega = \lfloor \cdot \rfloor_{1,\omega}$ whose restriction to $\mathbb{R}$ coincides with $\lfloor \cdot \rfloor$. Then, given $\alpha, \beta \in \mathbb{C}$, we put

$$\mathcal{B}_\omega(\alpha, \beta) = \{\lfloor n\alpha + \beta \rfloor_\omega : n \in \mathcal{O}\} \tag{1.10}$$

and call this the *Beatty-type set* associated with α, β and ω.

These sets admit a similar characterisation as the one furnished by Lemma 1.2 for ordinary Beatty sets (see Lemma 5.10 below). On combining this with results about the distribution of

$$p/\alpha \quad \mod \{\lambda_1 + \lambda_2\omega : \lambda_1, \lambda_2 \in [0, 1)\}$$

($\alpha \in \mathbb{C} \setminus \mathbb{K}$, p ranging over prime elements in $\mathcal{O}$) obtained in Chapter 4, we are able to detect infinitely many prime elements in $\mathcal{B}_\omega(\alpha, \beta)$ when $\alpha \in \mathbb{C} \setminus \mathbb{K}$ and $\mathbb{K}$ has class number 1 (see Theorem 5.5 below). Furthermore, using the Hurwitz continued fraction algorithm in $\mathbb{Z}[i]$, we provide an analogue of Theorem 1.6 for the Gaussian integers.

The work in Chapter 4 rests on the sieve method due to Harman, which we need in a formulation tailored to our particular setting; the proof of the sieve result itself is practically the one given in [35, 34], but we still include it in Appendix B and supporting results about quadratic extensions used throughout are collected in Appendix A. The fact that Harman's sieve can be applied to study the distribution of $p\vartheta \mod 1$ ($\vartheta \in \mathbb{R} \setminus \mathbb{Q}$) with p ranging over the rational primes is due to Harman himself. The adjustments necessary to study the distribution of

$$p\vartheta \quad \mod \{\lambda_1 + \lambda_2 i : \lambda_1, \lambda_2 \in [0, 1)\}$$

$(\vartheta \in \mathbb{C} \setminus \mathbb{Z}[i])$ with p ranging over prime elements in the ring of Gaussian integers $\mathbb{Z}[i]$ have been worked out by Baier [4]. The work in Chapter 4 is largely based on his paper.

We shall comment some more on the restriction to class number 1 present in Chapters 4 and 5. The reader will note that $\mathbb{K}$ having class number 1 means that its corresponding ring of integers $\mathcal{O}$ is a principal ideal domain. The celebrated *Baker–Heegner–Stark theorem* [41, 5, 77, 78] gives a complete list of these $\mathbb{K}$:

Theorem 1.7 (Baker–Heegner–Stark). *The imaginary quadratic number fields $\mathbb{K}$ with class number 1 are precisely those $\mathbb{Q}(\sqrt{d})$ with d from the list*

$$d = -1, -2, -3, -7, -11, -19, -43, -67, -163. \tag{1.11}$$

The restriction to class number 1 comes with the sieve method which makes use of unique factorisation into irreducible elements and $\mathcal{O}$, being a Dedekind domain, is a unique factorisation domain if and only if it is principal. Concerning the sieve method, one could get around this by sifting ideals (in fact, this is what is happening below the surface anyway), but the corresponding question for a bound for the smallest norm of prime ideals $\mathfrak{p} \subseteq \mathcal{O}$ ($\mathfrak{p} \neq \{0\}$) contained in $\mathcal{B}_\omega(\alpha, \beta)$ does not turn out to be meaningful, for the results of Chapters 4 and 5 can be used to show that $\mathcal{B}_\omega(\alpha, \beta)$ with $\alpha \in \mathbb{C} \setminus \mathbb{K}$ never fully contains a non-trivial prime ideal (at least when $\mathcal{O}$ has class number 1), and, likewise, asking for prime ideals $\mathfrak{p}$ containing $\mathcal{B}_\omega(\alpha, \beta)$ can also be shown to be the wrong type of question. On the other hand, Harman, Kumchev, and Lewis [32] investigate a setting in which one may associate elements $\xi_\mathfrak{a}$ of $\mathcal{O}$ to integral ideals $\mathfrak{a}$, belonging to some fixed element of the ideal class group of $\mathbb{K}$, in a more or less natural fashion; having done this, they sift for prime ideals $\mathfrak{p}$ amongst the set of said ideals $\mathfrak{a}$ with additional restrictions imposed on $\xi_\mathfrak{a}$. Imposing the restriction that $\xi_\mathfrak{a}$ be an element of some Beatty-type set $\mathcal{B}_\omega(\alpha, \beta)$ seems to be a problem worth investigating, but this is not implemented in this thesis.

On the contrary, we add that the restriction to class number 1 also brings some consolation, as it implies that all irreducible elements in $\mathcal{O}$ are automatically prime, thus sparing us the trouble of having to distinguish the two notions.

11

Chapter 2

Kloosterman sums with Beatty sequences

2.1 A problem concerning modular hyperbolas

In this section we shall be interested in the distribution of the points on *modular hyperbolas*

$$\mathcal{H}_{z,p} = \{(m, \tilde{m}) \in \mathbb{Z}^2 \cap [1, p)^2 : m\tilde{m} \equiv z \mod p\} \quad (p \nmid z), \tag{2.1}$$

where the letter p denotes a prime here and throughout. The interested reader is referred to [74] for a survey of questions related to this topic and various of their applications. Our particular point of departure shall be the following intriguing property of such hyperbolas:

Theorem 2.1. *For any prime p and z coprime to p there is always a point $(m, \tilde{m}) \in \mathcal{H}_{z,p}$ with $\max\{m, \tilde{m}\} \leq 2(\log p)p^{3/4}$.*

Proof. See, e.g., [39, p. 382].[2] □

Loosely speaking, the above theorem states that when sampling enough points $(1, ?), (2, ?'), \ldots \in \mathcal{H}_{z,p}$, at least one of the second coordinates will be not too large. The question we ask may be enunciated as follows:

Question 2.2. *If the first coordinate of $(m, \tilde{m})$ in Theorem 2.1 is additionally required to belong to a Beatty set with irrational slope, can one still prove a result like Theorem 2.1, that is, is it impossible for a Beatty set to contain only those elements $m < p$ with "large" corresponding $\tilde{m}$ from (2.1)?*

More specifically, for irrational $\alpha > 1$ and non-negative β, we shall be interested in solving

$$m\tilde{m} \equiv z \mod p \quad \text{with} \quad m \in \mathcal{B}(\alpha, \beta) \quad \text{and} \quad 1 \leq m, \tilde{m} < p \tag{2.2}$$

whilst keeping $\max\{m, \tilde{m}\}$ as small as possible. Thus, we shall want to bound

$$F(z; p) = \min\{\max\{m, \tilde{m}\} : (m, \tilde{m}) \text{ s.t. (2.2) holds}\}. \tag{2.3}$$

---

[2] The proof itself gives much more. Indeed, instead of restricting $(m, \tilde{m})$ to the square box $(0, 2(\log p)p^{3/4}]^2$, any rectangular box $\mathcal{R} \subseteq (0, p]^2$ with sufficiently large area can be seen to have a non-empty intersection with $\mathcal{H}_{z,p}$.

As an illustration, consider Fig. 4: therein, bounding $F(1;p)$ is equivalent to asking as to how large one must take the side length of a square with lower left corner positioned at $(0,0)$, in order to be guaranteed to find a black point inside. The smallest such square is sketched thick in the figure.

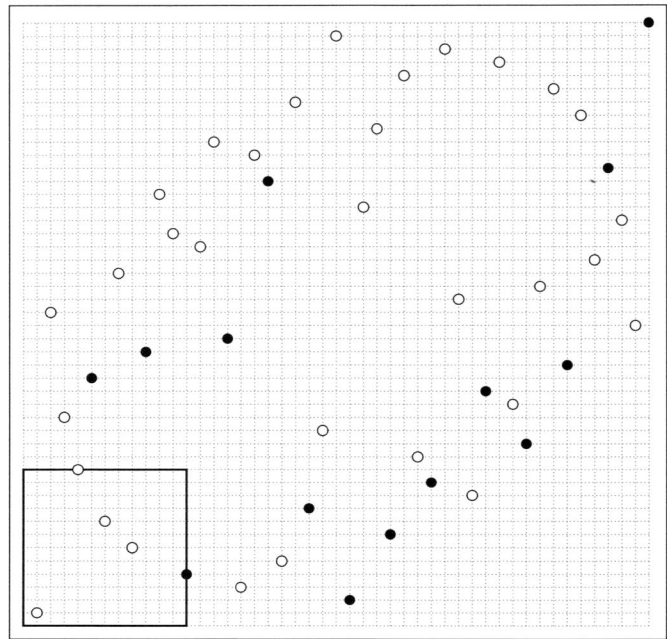

Figure 4: A grid showing the points $(m, \tilde{m}) \in \mathscr{H}_{1,p}$ where $p = 47$. If $m \in \mathcal{B}(\pi, e)$, then the point corresponding to $(m, \tilde{m})$ is filled.

We state our result, which answers *yes* to Question 2.2 in some instances, in Section 2.4. In the remainder of the present section we adapt the argument from [39, p. 382]: On writing

$$f_M(\tilde{m}) = \mathbf{1}_{\{\exists m \leq M \text{ s.t. (2.2) holds}\}}$$

one immediately observes that $F(z; p)$ from (2.3) may be written as

$$F(z; p) = \min\left\{ M < p : \sum_{\tilde{m} \leq M} f_M(\tilde{m}) > 0 \right\}.$$

From this one can proceed with a standard completing technique. Indeed, by [45, Lemma 12.1],

$$\left| \sum_{\tilde{m} \leq M} f_M(\tilde{m}) - \frac{M}{p} \sum_{\tilde{m} < p} f_M(\tilde{m}) \right| < (\log p) \max_{k < p} \left| \sum_{\tilde{m} < p} f_M(\tilde{m}) \, e_p(k\tilde{m}) \right|,$$

14

where $e_p(x) = \exp(2\pi i\, x/p)$. Here and below we shall write $\overline{m}$ for the unique integer in the range $[1, p)$ with $m\overline{m} \equiv 1 \bmod p$. Using this notation, we find

$$\sum_{\tilde{m} < p} f_M(\tilde{m})\, e_p(k\tilde{m}) = \sum_{\substack{\tilde{m} < p \\ m \in \mathcal{B}(\alpha,\beta) \\ m\tilde{m} \equiv z \bmod p}} \sum_{m \le M} e_p(k\tilde{m}) = \sum_{\substack{m \le M \\ m \in \mathcal{B}(\alpha,\beta)}} e_p(kz\overline{m}).$$

Thus, if $M < p$ is a positive integer such that

$$\frac{M(M - 1 - \beta - \alpha)}{p\alpha} \ge (\log p) \max_{y < p} \left| \sum_{\substack{m \le M \\ m \in \mathcal{B}(\alpha,\beta)}} e_p(y\overline{m}) \right|, \tag{2.4}$$

then $F(z; p) \le M$.

Therefore, aiming to be able to take M as small as possible, we are left with the task of giving a good upper bound for the sums on the right hand side of (2.4). Disregarding the restriction "$m \in \mathcal{B}(\alpha, \beta)$" for the moment and applying the completing technique yet again gives

$$\left| \sum_{m \le M} e_p(y\overline{m}) \right| \le (1 + \log p) \left| \max_{x \le p} \sum_{m < p} e_p(xm + y\overline{m}) \right|.$$

Of course, the sums on the right hand side are the well-known *Kloosterman sums*

$$S(x, y; p) = \sum_{m < p} e_p(xm + y\overline{m}), \tag{2.5}$$

introduced by Kloosterman [50] in his seminal refinement of the Hardy–Littlewood circle method to handle diagonal quadratic forms in four variables. He proved the bound

$$\left| \sum_{m < p} e_p(xm + y\overline{m}) \right| \ll p^{3/4} \gcd(x, y, p)^{1/4},$$

which was later improved by Weil [90] to

$$\left| \sum_{m < p} e_p(xm + y\overline{m}) \right| \le 2p^{1/2} \gcd(x, y, p)^{1/2}, \tag{2.6}$$

the latter being asymptotically optimal (see [45, Section 11.7]).

2.2 Main results

Returning to (2.4) and recalling the heuristic principle enunciated in Section 1.1.3, we might expect to be able to bound the right hand side of (2.4) non-trivially. However, the problem is more subtle, since the function

$$f(m) = \begin{cases} e_p(xm + y\overline{m}) & p \nmid m, \\ 0 & p \mid m, \end{cases} \tag{2.7}$$

15

for which we might want to compare (1.4) with (1.5) also depends on x, y and p, and we lack the necessary uniformity in those parameters. In particular, we would like to draw attention to the dependence on f of the estimate in Theorem 1.1 and indeed the proof given in [2] does not seem to yield what we lack at this point: as it stands, the argument proving the first estimate on page 87 of [2] only provides f-dependent implied constants.

Nonetheless, the main result of this chapter is that non-trivial bounds for

$$K_{\alpha,\beta}(x,y;p,N) = \sum_{\substack{n \leq N \\ p \nmid \lfloor n\alpha + \beta \rfloor}} e_p\big(x\lfloor n\alpha + \beta \rfloor + y\overline{\lfloor n\alpha + \beta \rfloor}\big) \tag{2.8}$$

can still be obtained. To state our results, let

$$D_{\alpha,\beta}(N) = \sup_{0 \leq x < y < 1} \left| \frac{\#\{n \leq N : \{n\alpha + \beta\} \in [x,y)\}}{N} - (y - x) \right| \tag{2.9}$$

denote the *discrepancy* of the finite sequence $(\{n\alpha + \beta\})_{n \leq N}$ and put $D_\alpha(N) = D_{\alpha,0}(N)$.

Theorem 2.3. *Suppose that p is a prime, $\beta \geq 0$ and $x, y \in \mathbb{Z}$ such that $p \nmid y$. Then, for any irrational $\alpha > 1$ and $N \leq p$, the sum $K_{\alpha,\beta}(x,y;p,N)$ given by (2.8) satisfies the bound*

$$|K_{\alpha,\beta}(x,y;p,N)| \ll_\alpha N^{297/512}p^{43/128} + ND_\alpha(N), \tag{2.10}$$

where the implied constant only depends on α.

The proof of this result is based on a method due to Banks and Shparlinski [10, 11, 12] which works if one can estimate certain exponential sums non-trivially. More precisely, we prove the following:

Proposition 2.4. *Suppose that p is a prime and $x, y \in \mathbb{Z}$ such that $p \nmid y$. If*

$$S(x,y,w;p) = \sum_{\substack{m < p \\ p \nmid (m+w)}} e_p(x\overline{m} + y\overline{m+w}),$$

then, for any set $\mathcal{W}$ containing only values that are distinct modulo p, we have

$$\sum_{w \in \mathcal{W}} |S(x,y,w;p)| \ll (\#\mathcal{W})^{3/4}p. \tag{2.11}$$

In the proof of Theorem 2.3, the proposition is applied with $x = -y$, thus becoming an estimate for the *periodic autocorrelation* of the finite sequence

$$0,\ e_p(y\overline{1}),\ e_p(y\overline{2}),\ \ldots,\ e_p(y\overline{p-1})$$

on average. Replacing the estimate (2.11) by a non-trivial estimate on an individual sum $|S(x,y,w;p)|$ appears to be an open problem (see also Remark 2.13).

2.3 Facts about the discrepancy $D_{\alpha,\beta}(N)$

In order to make use of Theorem 2.3, it is necessary to understand the discrepancy (2.9). As can be easily seen, in general, (2.9) depends on the shift β (consider, for instance, $(\alpha, N) = (\frac{15}{8}, 2)$ and $\beta \in \{0, \frac{1}{8}\}$). This dependency could be removed by defining the discrepancy as, e.g., Montgomery [65] does, but in any case we have the following result:

Lemma 2.5. *For any $\alpha, \beta \in \mathbb{R}$ and $N \in \mathbb{N}$, we have $D_{\alpha,\beta}(N) \leq 8D_\alpha(N)$.*

Proof. This follows easily from [53, Chapter 2, Theorem 1.3] and [65, Chapter 1, Eq. (8)]. $\qquad\square$

It is a classic result due to Bohl, Sierpiński and Weyl that the discrepancy of $(\{n\alpha\})_{n \leq N}$ tends to zero (see [91] and the references therein). Namely, we have:

Lemma 2.6 ([91, Satz 2]). *Let α be irrational. Then*

$$D_\alpha(N) = o_\alpha(1),$$

where the implied constant depends on α.

In particular, Theorem 2.3 is non-trivial for all irrational numbers $\alpha > 1$, provided that N and p are in the correct range (see discussion below).

For certain α, Lemma 2.6 can be sharpened considerably. To describe these numbers, recall that the *type* τ of an irrational number α is defined by

$$\tau = \sup\left\{\eta \in \mathbb{R} : \liminf_{q \to \infty} q^\eta \|\alpha q\| = 0\right\},$$

where $\|x\|$ denotes the distance of x to the nearest integer. (Note that Dirichlet's approximation theorem ensures that $\tau \geq 1$.)

Lemma 2.7 ([53, Chapter 2, Theorem 3.2]). *Let α be of finite type τ. Then, for every $\epsilon > 0$,*

$$D_\alpha(N) \ll_{\alpha,\epsilon} N^{-1/\tau+\epsilon},$$

where the implied constant only depends on α and ϵ.

Turning back to Theorem 2.3, we find that the first term on the right hand side of (2.10) is dominant, provided that the type of α is bounded away from $\frac{512}{43}$ (mind the assumption $N \leq p$). As an immediate consequence we may state:

Corollary 2.8. *Let $\epsilon > 0$. On the hypotheses of Theorem 2.3, and restricting to only those irrational $\alpha > 1$ of finite type $\leq \frac{512}{43} - \epsilon$, we have the bound*

$$|K_{\alpha,\beta}(x, y; p, N)| \ll_{\alpha,\epsilon} N^{297/512} p^{43/128}. \tag{2.12}$$

The famous result of Khintchine [49] asserts that almost all real numbers are of type 1 and the celebrated theorem of Roth [71] establishes that all real algebraic irrational numbers are of type 1; for numbers of finite type larger than 1, the Lebesgue measure fails to provide useful information about their abundance. However, by the Jarník–Besicovitch theorem [46, 21], we know that the Hausdorff dimension of all real numbers of type $> \tau$ is

$$\dim_H\{x \in \mathbb{R} \text{ of type} > \tau\} = \frac{2}{\tau + 1}.$$

Furthermore, observe that Theorem 2.3 is only non-trivial for p and N in some range of the type $N \leq p \ll_\epsilon N^{5/4-\epsilon}$.

2.4 Back to the original problem

We return to our initial intent of bounding (2.3). To this end, we put $M = \lfloor N\alpha + \beta \rfloor$ (with $N \geq 4$, say) in (2.4) and employ Corollary 2.8, tacitly assuming its hypotheses in the process. Then, after a short calculation, we find that (2.4) is satisfied provided that

$$(\log p)N^{-2+297/512}p^{1+43/128} \leq C(\alpha, \epsilon) \tag{2.13}$$

with some $C(\alpha, \epsilon) > 0$ depending only on α and ϵ. Letting

$$N = \lfloor p^{684/727} \log p \rfloor,$$

the condition (2.13) will be satisfied for p sufficiently large in terms of α and ϵ.

Theorem 2.9. *Let $\epsilon > 0$ and β be non-negative. Moreover, assume that $\alpha > 1$ is irrational and of finite type $\leq \frac{512}{43} - \epsilon$. Then there is a number $p_0(\alpha, \epsilon)$ such that, for all primes $p \geq p_0(\alpha, \epsilon)$ and z coprime to p, there is a point $(m, \tilde{m}) \in \mathcal{H}_{z,p}$ with $m \in \mathcal{B}(\alpha, \beta)$ and*

$$\max\{m, \tilde{m}\} \leq \alpha p^{684/727} \log p + \beta. \tag{2.14}$$

Remark 2.10. *If $(m, \tilde{m})$ is a solution of (2.2), then*

$$m \geq \min \mathcal{B}(\alpha, \beta) = \lfloor \alpha + \beta \rfloor > \beta.$$

Hence, the dependence of the right hand side of (2.14) on β is not a deficiency in the proof, as it is inherent to the problem.

2.5 Proofs

2.5.1 Proof of Proposition 2.4

We start by deducing Proposition 2.4 from the following result:

Proposition 2.11. *On assuming the hypotheses of Proposition 2.4, we have*

$$\sum_{w \in \mathscr{W}} |S(x, y, w; p)|^4 \ll p^4. \tag{2.15}$$

Proof of Proposition 2.4. Note that, by Hölder's inequality, we have

$$\sum_{w \in \mathscr{W}} |S(x, y, w; p)| \leq (\#\mathscr{W})^{3/4} \cdot \left(\sum_{w \in \mathscr{W}} |S(x, y, w; p)|^4 \right)^{1/4}. \tag{2.16}$$

The result now follows after applying Proposition 2.11. $\qquad\square$

To prove Proposition 2.11, we adapt Kloosterman's original argument for bounding his sums (see [50, Section 2.43] and the comments made in Remark 2.13 below). The argument is based on using a transformation property of the sums $S(x, y, w; p)$ (see (2.24)) in order to find a large contribution of $|S(x, y, w; p)|^4$ in an average over the first three parameters. The average is then seen to count the number of solutions to certain congruence equations, and this number can be bounded non-trivially.

We start with an analysis of the congruence equations in question. To this end, write $\boldsymbol{m} = (m_1, m_2, m_3, m_4)$ and let

$$A(\boldsymbol{m}, u) = \sum_{k \leq 4} (-1)^k \overline{m_k + u} \tag{2.17}$$

if $p \nmid (m_1 + u)(m_2 + u)(m_3 + u)(m_4 + u)$ and, for notational simplicity, $A(\boldsymbol{m}, u) = 1$ otherwise. The key result is the following:

Lemma 2.12. *Let p be a prime and $A(\boldsymbol{m}, u)$ be given by (2.17) and suppose that*

$$\mathscr{X}_u = \{ \boldsymbol{m} \in \{1, \ldots, p\}^4 : A(\boldsymbol{m}, 0) \equiv A(\boldsymbol{m}, u) \equiv 0 \bmod p \}. \tag{2.18}$$

Then

$$\sum_{u \leq p} \#\mathscr{X}_u \ll p^3. \tag{2.19}$$

Proof. Consider $\boldsymbol{m} = (m_1, m_2, m_3, m_4) \in \mathscr{X}_u$. Certainly we have

$$p \nmid \prod_{k \leq 4} m_k(m_k + u). \tag{2.20}$$

Additionally, from $p \mid A(\boldsymbol{m}, 0)$ and $p \mid A(\boldsymbol{m}, u)$, we deduce that

$$\begin{cases} \overline{m_1} + \overline{m_3} \equiv \overline{m_2} + \overline{m_4} & \bmod p, \\ \overline{m_1 + u} + \overline{m_3 + u} \equiv \overline{m_2 + u} + \overline{m_4 + u} & \bmod p. \end{cases} \tag{2.21}$$

This is satisfied trivially if (m_1, m_3) is a permutation of (m_2, m_4). There are at most $2(p-1)^2$ such trivial solutions. Assume next that $\boldsymbol{m}$ is a non-trivial solution and,

19

additionally, that neither expression in (2.21) is $\equiv 0$; this additional assumption excludes at most $2(p-1)^2$ values of m. From (2.21) it follows that

$$(\overline{m_1} + \overline{m_3})m_1 m_3 (m_2 + m_4)$$
$$\equiv (m_2 + m_4)(m_1 + m_3)$$
$$\equiv (m_1 + m_3)m_2 m_4 (\overline{m_2} + \overline{m_4}) \quad \mod p.$$

Upon cancelling $\overline{m_1} + \overline{m_3} \equiv \overline{m_2} + \overline{m_4} \not\equiv 0 \mod p$, we find that

$$m_1 m_3 (m_2 + m_4) \equiv (m_1 + m_3)m_2 m_4 \quad \mod p,$$

which in turn may be rearranged to

$$m_1(m_2 m_3 + m_3 m_4 - m_2 m_4) \equiv m_2 m_3 m_4 \quad \mod p. \tag{2.22}$$

Note that the term $m_2 m_3 + m_3 m_4 - m_2 m_4 \mod p$ cannot vanish, for otherwise (2.22) implies $p \mid m_2 m_3 m_4$, in contradiction to (2.20). Hence, for any m satisfying the above assumptions, we may compute $m_1 \mod p$ from (m_2, m_3, m_4). Indeed, by (2.22),

$$m_1 \equiv m_2 m_3 m_4 \overline{(m_2 m_3 + m_3 m_4 - m_2 m_4)} \quad \mod p.$$

Next, we claim that m belongs to at most two of the sets $\mathscr{X}_u$ ($u = 1, \ldots, p$). To see this, first observe that, along similar lines as the deduction of (2.22) from (2.21), we have

$$(m_1 + u)(m_3 + u)(m_2 + m_4 + 2u)$$
$$\equiv (m_1 + m_3 + 2u)(m_2 + u)(m_4 + u) \quad \mod p. \tag{2.23}$$

Recalling that m was assumed to be non-trivial, the difference of the left and right hand side of the above equation is seen to be a *non-zero* polynomial of degree at most two in u. Since $\mathbb{Z}/p\mathbb{Z}$ is an integral domain, the claim follows. From this, and taking the trivial solutions into account, we conclude that

$$\sum_{u<p} \#\mathscr{X}_u \leq p(2p^2 + 4(p-1)^2) \ll p^3. \qquad \qquad \square$$

Proof of Proposition 2.11. First observe that

$$S(x, y, w; p) = S(ax, ay, aw; p) \tag{2.24}$$

for any a indivisible by p. Thus, the average

$$\Sigma = \sum_{r \leq p} \sum_{t \leq p} \sum_{u < p} |S(r, t, u; p)|^4$$

contains $p-1$ copies of $|S(x, y, w; p)|^4$. Hence,

$$(p-1) \sum_{w \in \mathscr{W}} |S(x, y, w; p)|^4 \leq \Sigma. \tag{2.25}$$

To estimate Σ, write $\boldsymbol{m} = (m_1, m_2, m_3, m_4)$ and start by squaring out each term $|S(r, t, u; p)|^4$ twice to obtain

$$\Sigma = \sum_{r \leq p} \sum_{t \leq p} \sum_{u < p} \sum_{\substack{m_1 < p}} \sum_{\substack{m_2 < p}} \sum_{\substack{m_3 < p}} \sum_{\substack{m_4 < p \\ p \nmid (m_1+u)(m_2+u)(m_3+u)(m_4+u)}} e_p(rA(\boldsymbol{m}, 0) + tA(\boldsymbol{m}, u)),$$

where $A(\boldsymbol{m}, u)$ is given by (2.17).

After moving the summation over r and t to the right and observing that

$$\sum_{r \leq p} e_p(rx) = \begin{cases} p & \text{if } p \mid x, \\ 0 & \text{if } p \nmid x, \end{cases}$$

we find that

$$\Sigma = p^2 \sum_{u < p} \# \mathscr{X}_u,$$

where $\mathscr{X}_u$ is given by (2.18). By Lemma 2.12, $\Sigma \ll p^5$, so that from (2.25) we conclude (2.15). □

Remark 2.13. *It should be stressed that in the proof of Proposition 2.11 we do not obtain a non-trivial bound for an individual sum $S(x, y, w; p)$; it is only the averaging over $w \in \mathscr{W}$ which makes this result non-trivial. This may be compared with Kloosterman [50], who obtains*

$$|S(x, y; p)| \ll p^{3/4}(x, y, p)^{1/4},$$

(recall (2.5)) by exploiting the transformation property

$$S(x, y; p) = S(ax, \bar{a}y; p) \quad (p \nmid a)$$

in place of (2.24). Notice that only two parameters are transformed, so that when considering the obvious analogue of Σ one only has to average over two parameters; the saving obtained from the resulting congruence equations having few solutions thus has a larger impact.

2.5.2 Proof of Theorem 2.3

We shall need the following result which follows easily from the pigeonhole principle:

Lemma 2.14 ([10, Lemma 3.3]). *Let α be a fixed irrational number. Then, for every positive integer K and real number $\Delta \in (0, 1]$, there exists a real number γ such that*

$$\#\{k \leq K : \{k\alpha + \gamma\} < \Delta\} \geq 0.5 \, K\Delta.$$

Now let $K \leq N$ be a positive integer and $\Delta \in (0, 1)$ to be determined later (see (2.32) below). Then, by Lemma 2.14, there is some real number γ such that the set

$$\mathscr{K} = \{k \leq K : \{k\alpha + \gamma\} < \Delta\} \tag{2.26}$$

has cardinality
$$\#\mathcal{K} \geq 0.5\, K\Delta. \tag{2.27}$$

Furthermore, let
$$\mathcal{N} = \{1 \leq n \leq N : \{n\alpha + \beta - \gamma\} < 1 - \Delta\} \tag{2.28}$$

and $\mathcal{N}^{\mathrm{c}} = \{1, \ldots, N\} \setminus \mathcal{N}$. Clearly, recalling (2.9),
$$\#\mathcal{N}^{\mathrm{c}} = N\Delta + O(N D_{\alpha,\beta}(N)). \tag{2.29}$$

Now, writing
$$x_{n,k} = e_p\big(x\lfloor (n+k)\alpha + \beta \rfloor + y\overline{\lfloor (n+k)\alpha + \beta \rfloor}\big)$$

for the moment, for every $k \in \mathcal{K}$, we find that
$$
\begin{aligned}
K_{\alpha,\beta}(x, y; p, N) &= \sum_{\substack{n \leq N \\ p \nmid \lfloor (n+k)\alpha + \beta \rfloor}} x_{n,k} + O(k) \\
&= \sum_{\substack{n \leq N \\ p \nmid \lfloor (n+k)\alpha + \beta \rfloor}} x_{n,k} + O(K) \\
&= \sum_{\substack{n \in \mathcal{N} \\ p \nmid \lfloor (n+k)\alpha + \beta \rfloor}} x_{n,k} + O(K + \#\mathcal{N}^{\mathrm{c}}).
\end{aligned}
$$

Consequently,
$$K_{\alpha,\beta}(x, y; p, N) = \frac{W}{\#\mathcal{K}} + O(K + \#\mathcal{N}^{\mathrm{c}}), \tag{2.30}$$

where
$$W = \sum_{\substack{n \in \mathcal{N} \\ p \nmid \lfloor (n+k)\alpha + \beta \rfloor}} \sum_{k \in \mathcal{K}} e_p\big(x\lfloor (n+k)\alpha + \beta \rfloor + y\overline{\lfloor (n+k)\alpha + \beta \rfloor}\big).$$

For any $(n, k) \in \mathcal{N} \times \mathcal{K}$, a simple calculation shows that
$$\lfloor (n+k)\alpha + \beta \rfloor = \lfloor n\alpha + \beta - \gamma \rfloor + \lfloor k\alpha + \gamma \rfloor.$$

Next, we apply Cauchy's inequality to W, getting
$$
\begin{aligned}
|W|^2 &\leq \#\mathcal{N} \cdot \sum_{n \in \mathcal{N}} \left| \sum_{\substack{k \in \mathcal{K} \\ p \nmid \lfloor (n+k)\alpha + \beta \rfloor}} e_p\big(x\lfloor (n+k)\alpha + \beta \rfloor + y\overline{\lfloor (n+k)\alpha + \beta \rfloor}\big) \right|^2 \\
&\ll_\alpha N \cdot \sum_{s \leq p} \left| \sum_{\substack{k \in \mathcal{K} \\ p \nmid (s + \lfloor k\alpha + \gamma \rfloor)}} e_p\big(x(s + \lfloor k\alpha + \gamma \rfloor) + y\overline{(s + \lfloor k\alpha + \gamma \rfloor)}\big) \right|^2,
\end{aligned}
$$

where replacing the summation over $n \in \mathcal{N}$ by the summation over $s \leq p$ is allowed, since our assumption $N \leq p$ ensures that

$$\#\{n \in \mathcal{N} : \lfloor n\alpha + \beta - \gamma \rfloor \equiv s \bmod p\} < 1 + \alpha \ll_\alpha 1.$$

On squaring out the inner sum, we find that

$$|W|^2 \ll_\alpha N \cdot \sum_{\substack{s \leq p, \, k, \ell \in \mathcal{K} \\ p \nmid (s + \lfloor k\alpha + \gamma \rfloor) \\ p \nmid (s + \lfloor \ell\alpha + \gamma \rfloor)}} \mathrm{e}_p(x(\lfloor k\alpha + \gamma \rfloor + \lfloor \ell\alpha + \gamma \rfloor)) \times$$
$$\times \, \mathrm{e}_p\big(y(\overline{s + \lfloor k\alpha + \gamma \rfloor} - \overline{s + \lfloor \ell\alpha + \gamma \rfloor})\big).$$

This is

$$\ll_\alpha N \cdot \sum_{k, \ell \in \mathcal{K}} \left| \sum_{\substack{s \leq p \\ p \nmid (s + \lfloor k\alpha + \gamma \rfloor) \\ p \nmid (s + \lfloor \ell\alpha + \gamma \rfloor)}} \mathrm{e}_p\big(y(\overline{s + \lfloor k\alpha + \gamma \rfloor} - \overline{s + \lfloor \ell\alpha + \gamma \rfloor})\big) \right|. \qquad (2.31)$$

Upon writing

$$\tilde{\mathcal{K}} = \{\lfloor k\alpha + \gamma \rfloor - \lfloor \ell\alpha + \gamma \rfloor : k, \ell \in \mathcal{K}\},$$

we infer

$$|W|^2 \ll_\alpha N \cdot \#\mathcal{K} \sum_{w \in \tilde{\mathcal{K}}} \left| \sum_{\substack{s < p \\ p \nmid (s + w)}} \mathrm{e}_p(y(\overline{s} - \overline{s + w})) \right|.$$

Hence, by Proposition 2.4,

$$|W|^2 \ll_\alpha N \cdot (\#\mathcal{K})^{7/4} \cdot p.$$

In view of (2.30), we find that

$$|K_{\alpha,\beta}(x, y; p, N)| \ll_\alpha \sqrt{\frac{Np}{(\#\mathcal{K})^{1/4}} + K + \#\mathcal{N}^{\mathrm{c}}}.$$

Consequently, upon gathering (2.27), (2.29) and Lemma 2.5, we obtain the bound

$$|K_{\alpha,\beta}(x, y; p, N)| \ll_\alpha (Np)^{1/2}(K\Delta)^{-1/8} + K + N\Delta + ND_\alpha(N).$$

With the choices

$$\Delta = N^{-105/128} p^{21/32} \quad \text{and} \quad K = \lceil N\Delta \rceil \qquad (2.32)$$

we then conclude the assertion of the theorem provided that $\Delta < 1$, i.e., when $\log p / \log N < \frac{5}{4}$. On the other hand, if $\log p / \log N \geq \frac{5}{4}$, then the theorem asserts nothing more than the trivial bound

$$|K_{\alpha,\beta}(x, y; p, N)| \leq N,$$

so the proof is complete.

Remark 2.15. *The reader aware of the Three Gap Theorem mentioned in Chapter 1 may ponder whether one can do without the discrepancy in the above argument. Certainly the distribution of the points $\{k\alpha + \gamma\}$ and $\{n\alpha + \beta - \gamma\}$ in Eqs. (2.26) and (2.28) respectively admits more structure than we have effectively used here, so one may naively hope to obtain some discrepancy-free substitute for (2.29), maybe choosing γ more carefully, but still retaining some bound of the type (2.27). However, a closer inspection of the situation quickly shatters these expectations, because one of the gaps in the Three Gap Theorem can be quite small (this may be deduced, e.g., from the comment below Theorem 2.2 in [85] and [85, Theorem 3.3]), thus allowing the points $\{n\alpha\}$, $n = 1, 2, \ldots$, to form clusters, opening up the possibility for $\#\mathcal{N}^c$ to be inconveniently large.*

Chapter 3

Generalised Beatty sets

3.1 Introduction

Recall our definition (1.9) of what we called *generalised Beatty sets* in Chapter 1, namely

$$\mathcal{B}(\boldsymbol{\alpha}, \beta) = \{\lfloor m\alpha_1 + n\alpha_2 + \beta \rfloor : m, n \in \mathbb{N}\},$$

where $\boldsymbol{\alpha} = (\alpha_1, \alpha_2)$ is a real vector with coordinates ≥ 1 and β is some non-negative real number. As we have already discussed there, our motive in introducing them was to define sets similar to Beatty sets for which a stronger form of Theorem 1.6 could be obtained.

For instance, a quick calculation gives

$$\mathcal{B}((\sqrt{2}, \pi), 0) = \{4, 5, 7, 8, 9, 10, 11, 12, 13, 14, 15, 16, \ldots\},$$

and one might eagerly note that lots of primes show up here. However, after calculating even more elements, one cannot help but suspect that

$$\mathcal{B}((\sqrt{2}, \pi), 0) = \mathbb{N} \setminus \{1, 2, 3, 6\}. \tag{3.1}$$

In fact, upon noting that $\mathbb{N} = \mathcal{B}(1, 0)$ is a Beatty set, the above equation takes the form

$$\mathcal{B}(\boldsymbol{\alpha}, \beta) = \{\text{some Beatty set}\} \setminus \{\text{finitely many exceptions}\}. \tag{3.2}$$

Perhaps surprisingly, this is the true nature of things here and our results in this chapter may be described as follows:

1. We show that (3.2) always holds.

2. We exhibit a suitable choice for the Beatty set on the right hand side of (3.2).

3. We give bounds for the largest exception in (3.2).

3.2 Statement of results

A set $\mathcal{X} \subseteq \mathbb{Z}$ is said to be *cofinite* with respect to $\mathcal{Y} \subseteq \mathbb{Z}$ if $\mathcal{X} \subseteq \mathcal{Y}$ and the set $\mathcal{X}^c = \mathcal{Y} \setminus \mathcal{X}$ is finite. In this case we write $\operatorname{cosup}_{\mathcal{Y}} \mathcal{X}$ for the supremum of $\mathcal{X}^c$. In particular, $\operatorname{cosup}_{\mathcal{Y}} \mathcal{Y} = -\infty$. If we drop the reference to $\mathcal{Y}$, then it is understood that $\mathcal{Y} = \mathbb{N}$. Here and throughout, $\boldsymbol{\alpha}$ always denotes a vector (α_1, α_2) with real coordinates ≥ 1.

3.2.1 Irrational α_1/α_2

Theorem 3.1. *Let $\alpha_1, \alpha_2 \geq 1$ and $\beta \geq 0$ be real numbers and suppose that the quotient α_2/α_1 is irrational. Then $\mathcal{B}(\boldsymbol{\alpha}, \beta)$ is cofinite.*

To state a bound on $\operatorname{cosup} \mathcal{B}(\boldsymbol{\alpha}, \beta)$, we need some vocabulary: for a real number α write

$$\alpha = [x_0; x_1, x_2, \ldots] = x_0 + \cfrac{1}{x_1 + \cfrac{1}{x_2 + \cfrac{1}{\ddots}}} \tag{3.3}$$

for its *(regular) continued fraction expansion*. If α is irrational, then this expansion has infinitely many terms and the so-called *partial quotients* x_j $(j = 0, 1, 2, \ldots)$ are uniquely determined. Any finite cutoff

$$[x_0; x_1, x_2, \ldots, x_r]$$

may be written as a reduced fraction $\frac{a}{q}$, the *rth convergent* to α. (Mind that we start with the 0th convergent.) For more background on continued fractions the reader is referred to [67, 31].

Proposition 3.2. *Assume the hypotheses of Theorem 3.1, and let $\frac{a}{q}$ be the rth convergent to α_2/α_1, where*

$$r = 1 + \lceil \log(2\alpha_1)/\log G \rceil \tag{3.4}$$

and $G = \frac{1}{2}(1 + \sqrt{5})$ is the Golden ratio. Then

$$\operatorname{cosup} \mathcal{B}(\boldsymbol{\alpha}, \beta) < \alpha_1 + q\alpha_2 + \beta. \tag{3.5}$$

To illustrate this bound, we note that it can be used to prove (3.1) which was previously stated in the introduction without proof. Indeed, a calculation reveals that $r = 4$ in (3.4) and

$$\pi/\sqrt{2} = [2; 4, 1, 1, 15, \ldots].$$

This leads to the fraction $\frac{a}{q} = [2; 4, 1, 1, 15] = \frac{311}{140}$. Thus,

$$\operatorname{cosup} \mathcal{B}((\sqrt{2}, \pi), 0) < \sqrt{2} + 140\pi \approx 441.2,$$

and a calculation of all small elements in $\mathcal{B}((\sqrt{2}, \pi), 0)$ now establishes (3.1). We note that when reversing the roles of π and $\sqrt{2}$ (which incidentally does not change the associated Beatty set) one obtains the bound

$$\operatorname{cosup} \mathcal{B}((\pi, \sqrt{2}), 0) < \pi + 311\sqrt{2} \approx 442.962.$$

3.2.2 Rational coordinates

If α has rational coordinates, then $\mathcal{B}(\alpha, \beta)$ can be described quite concisely:

Theorem 3.3. *Let $\alpha_1 = \frac{a_1}{q_1}$ and $\alpha_2 = \frac{a_2}{q_2}$ be reduced fractions with $\alpha_1, \alpha_2 \geq 1$. Furthermore, let β be a non-negative real number and put*

$$\tilde{\alpha} = \frac{c}{q_1 q_2}, \quad \tilde{\beta} = \alpha_1 + \alpha_2 - \tilde{\alpha} + \beta, \tag{3.6}$$

where c denotes the greatest common divisor of $a_1 q_2$ and $a_2 q_1$. Then $\mathcal{B}(\alpha, \beta)$ is cofinite with respect to $\mathcal{B}(\tilde{\alpha}, \tilde{\beta})$.

Note that Beatty sets $\mathcal{B}(\tilde{\alpha}, \tilde{\beta})$ with rational slope $\tilde{\alpha}$ can be written as a union of arithmetic progressions. Moreover, we have a simple analogue of Proposition 3.2:

Corollary 3.4. *On the hypotheses of Theorem 3.3, the set $\mathcal{B}(\alpha, \beta)$ is cofinite if and only if*

$$c \leq q_1 q_2. \tag{3.7}$$

Moreover, if (3.7) holds, then $\operatorname{cosup} \mathcal{B}(\alpha, \beta) \leq \lfloor a_1 a_2/c + \beta \rfloor$.

3.2.3 The residual case

For the sake of giving a complete treatment of all possible cases we also include the following theorem, although it does not differ greatly from the previous one.

Theorem 3.5. *Let $\alpha_1 > 1$ be irrational and $\alpha_2 = \frac{a}{q}\alpha_1$ with some reduced fraction $\frac{a}{q}$. Furthermore, let β be a non-negative real number and put*

$$\tilde{\alpha} = \frac{\alpha_1}{q} = \frac{\alpha_2}{a}, \quad \tilde{\beta} = \tilde{\alpha} + \beta. \tag{3.8}$$

Then $\mathcal{B}(\alpha, \beta)$ is cofinite with respect to $\mathcal{B}(\tilde{\alpha}, \tilde{\beta})$.

A result corresponding to Corollary 3.4 is also immediate.

3.3 Proof of the results for irrational α_1/α_2

Before turning to the details, we sketch the underlying idea. We note that $\mathcal{B}(\alpha, \beta)$ is a union of ordinary Beatty sets, namely

$$\mathcal{B}(\alpha, \beta) = \bigcup_{n \geq 1} \mathcal{B}(\alpha_1, n\alpha_2 + \beta). \tag{3.9}$$

By Lemma 1.2 membership of $x \in \mathbb{Z}$ to each of these Beatty sets is determined by the fractional part of x/α_k ($k = 1, 2$) belonging to a certain interval modulo one and x not being too small. On varying n, these "detector intervals" can be shown to eventually cover $\mathbb{R}/\mathbb{Z}$ (here the assumption that α_1/α_2 be irrational enters the picture) and, consequently, there is a finite collection of Beatty sets $\mathcal{B}(\alpha_1, n\alpha_2 + \beta)$ such that every

27

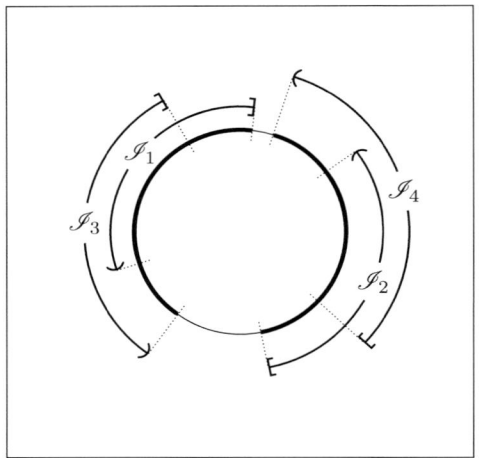 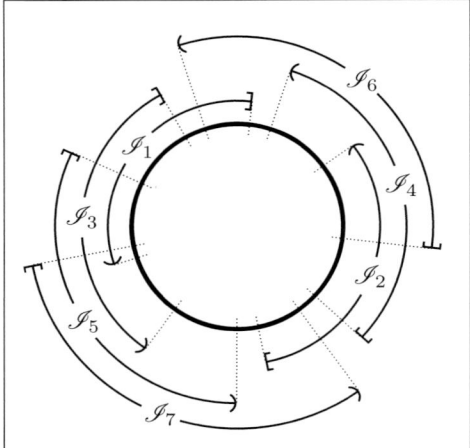

Figure 5: Intervals $\mathscr{I}_n$ modulo one (identified with arcs on a circle and lifted up for the sake of readability) "detecting" elements of $\mathcal{B}(\pi, n\sqrt{2})$ $(n = 1, 2, 3, 4)$. When drawing the picture for $n = 1, 2, \ldots, 7$, the full circle is covered.

sufficiently large integer x belongs to at least one element of said collection. The reader is referred to Fig. 5, which illustrates this point for the generalised Beatty set in (3.1).

Suppose that α has the continued fraction expansion (3.3) and let $\frac{a}{q}$ be the rth convergent to α. Then

$$|q\alpha - a| \leq q^{-1}. \tag{3.10}$$

The size of a and q depends, of course, on the values of $x_0, \ldots, x_r$. However, via comparison with the continued fraction expansion of the Golden ratio,

$$G = \tfrac{1}{2}(1 + \sqrt{5}) = [1; 1, 1, \ldots], \tag{3.11}$$

and using Binet's formula, a lower bound in terms of r can be given (see, e.g., [79] for the details).

Lemma 3.6. *Let $\frac{a}{q}$ be the rth convergent to some real number $\alpha > 0$. Then $q \geq G^{r-1}$, where G is given by (3.11).*

We shall require some very basic distribution properties of the sequence $(n\alpha)_n$ modulo one; of course, much more is true (see, e.g., [53] or the discussion of the discrepancy $D_{\alpha,\beta}(N)$ in Chapter 2), but the following simple lemma suffices for our purpose.

Lemma 3.7. *Let α be a real number and suppose that $\frac{a}{q}$ is a reduced fraction satisfying (3.10). Then*

$$\sup_{\xi \in [0,1)} \min_{n \leq q} \{\xi - n\alpha\} \leq 2q^{-1},$$

where $\{\xi\} = \xi - \lfloor \xi \rfloor$ denotes the fractional part of ξ.

Proof. This is immediate from the triangle inequality. $\qquad\square$

Proof of Theorem 3.1. Let x be large, $x > x_0$ say, where the value of x_0 is yet to be determined. We intend to show that $x \in \mathcal{B}(\boldsymbol{\alpha}, \beta)$. By (3.9) it suffices to exhibit some n such that $x \in \mathcal{B}(\alpha_1, n\alpha_2 + \beta)$. In view of Lemma 1.2, this is equivalent to

$$x\alpha_1^{-1} - n\alpha_1^{-1}\alpha_2 \in \left[\frac{\beta - 1}{\alpha_1}, \frac{\beta}{\alpha_1}\right) \mod 1, \qquad (3.12)$$

$$n < (x + 1 - \beta - \alpha_1)\alpha_2^{-1}. \qquad (3.13)$$

Now let $\frac{a}{q}$ be the rth convergent to $\alpha = \alpha_1^{-1}\alpha_2$. By taking r sufficiently large, we may assume that

$$q > 2\alpha_1. \qquad (3.14)$$

Then Lemma 3.7 ensures the existence of some $n \le q$ satisfying (3.12). The theorem now follows by choosing x_0 such that the right hand side of (3.13) is larger than q. $\qquad\square$

To give a *proof of Proposition 3.2,* a closer inspection of the above proof suffices: on applying Lemma 3.6, we find that (3.14) is satisfied if

$$1 + \log(2\alpha_1)/\log G \le r.$$

Since $\operatorname{cosup}\mathcal{B}(\boldsymbol{\alpha}, \beta) \le x_0$, we arrive at (3.5) after determining an admissible value for x_0 in the way outlined above. $\qquad\square$

Remark 3.8. *Of course, given some α for which the growth of the denominators of its rth convergents as $r \to \infty$ is known, Proposition 3.2 may be replaced with some smaller quantity.*

3.4 The case when α_1/α_2 is rational

As a motivation recall that given an integer vector $\boldsymbol{x} \in \mathbb{Z}^r$ $(r \ge 2)$ with positive coprime coordinates, one has

$$\mathbb{Z}x_1 + \ldots + \mathbb{Z}x_r = \mathbb{Z}.$$

When one replaces $\mathbb{Z}$ with $\mathbb{N}_0$, the set of non-negative integers, on the left hand side, then every sufficiently large integer on the right hand side can still be represented. The largest integer that cannot be thus represented is called the *Frobenius number* of $\boldsymbol{x}$ and shall be denoted by $g(\boldsymbol{x})$, i.e.,

$$g(\boldsymbol{x}) = \operatorname{cosup}_{\mathbb{N}_0}(\mathbb{N}_0 x_1 + \ldots + \mathbb{N}_0 x_r).$$

If $r = 2$, then one has the simple formula

$$g(x_1, x_2) = x_1 x_2 - x_1 - x_2.$$

However, to fit our convention that Beatty sets are generated via multiplication by *positive* integers, we shall work with

$$g_+(x_1, x_2) = \text{cosup}(\mathbb{N}x_1 + \mathbb{N}x_2) = x_1 x_2. \qquad (3.15)$$

Now looking at a Beatty set $\mathcal{B}(\boldsymbol{x}, \beta)$, one finds that

$$\mathcal{B}(\boldsymbol{x}, \beta) \supseteq \{n \in \mathbb{N} : n > g_+(x_1, x_2) + \beta\}.$$

It is a straightforward matter to extend this argument:

Proof of Theorem 3.3. For positive integers m, n write

$$m\frac{a_1}{q_1} + n\frac{a_2}{q_2} = \left(m\frac{a_1 q_2}{c} + n\frac{a_2 q_1}{c}\right)\frac{c}{q_1 q_2}. \qquad (3.16)$$

This already proves the inclusion $\mathcal{B}(\alpha, \beta) \subseteq \mathcal{B}(\tilde{\alpha}, \tilde{\beta})$. Since the expression in the parentheses in (3.16) can be made to equal any integer

$$> g_+(a_1 q_2 c^{-1}, a_2 q_1 c^{-1}) = \ell \quad \text{(say)}$$

if the positive integers m and n are chosen appropriately, we have

$$\text{cosup}_{\mathcal{B}(\tilde{\alpha}, \tilde{\beta})} \mathcal{B}(\alpha, \beta) \leq \left\lfloor \frac{\ell c}{q_1 q_2} + \beta \right\rfloor = \left\lfloor \frac{a_1 a_2}{c} + \beta \right\rfloor. \qquad (3.17)$$

This proves the theorem. □

Proof of Corollary 3.4. This is a trivial consequence of Theorem 3.3, and the fact that $\mathcal{B}(\tilde{\alpha}, \tilde{\beta})$ is cofinite if and only if $\tilde{\alpha} \leq 1$; the stated bound is obtained from (3.17) upon noting that $\mathcal{B}(\tilde{\alpha}, \tilde{\beta}) = \{x \in \mathbb{N} : x > \alpha + \tilde{\beta} - 1\}$ if (3.7) holds. □

In the setting of Corollary 3.4, one may naively suspect that the inequality given in (3.17) may in fact be an equality. However, this is not generally the case. Indeed, when $\tilde{\alpha} \leq \frac{1}{2}$, the problem of determining $\text{cosup}\,\mathcal{B}(\boldsymbol{\alpha}, \beta)$ is related to the position of *large gaps* (with respect to $\tilde{\alpha}$) in

$$\mathbb{N}a_1 q_2 c^{-1} + \mathbb{N}a_2 q_1 c^{-1}. \qquad (3.18)$$

To illustrate this point, consider the generalised Beatty set

$$\mathscr{B} = \mathcal{B}((\tfrac{3}{2}, 2), 0) = \{3, 5, 6, 7, 8, \ldots\} = \mathbb{N} \setminus \{1, 2, 4\}.$$

Here (3.18) takes the form

$$\mathbb{N}3 + \mathbb{N}4 = \mathbb{N} \setminus \{1, 2, 3, 4, 5, 6, \_\, 8, 9, \_\_\, 12\}$$

and we have $\tilde{\alpha} = \frac{1}{2}$. Note that $g_+(4, 3) = 12$, yet

$$\text{cosup}\,\mathscr{B} < 6 = \lfloor 12\tilde{\alpha} \rfloor,$$

because $\lfloor 12\tilde{\alpha} \rfloor = \lfloor 13\tilde{\alpha} \rfloor$ and 13 is an element of $\mathbb{N}3 + \mathbb{N}4$. However,

$$\text{cosup}\,\mathscr{B} = 4 = \lfloor 8\tilde{\alpha} \rfloor = \lfloor 9\tilde{\alpha} \rfloor.$$

Proof of Theorem 3.5. Simply write

$$m\alpha_1 + n\frac{a}{q}\alpha_1 = (mq + na)\frac{\alpha_1}{q}$$

and argue as in the proof of Theorem 3.3. $\square$

Chapter 4

Diophantine approximation with primes from $\mathcal{O}$

The material in the present chapter was worked out with the intent of detecting prime elements in the Beatty-type sets $\mathcal{B}_\omega(\alpha, \beta)$ and thereby establishing an analogue of Theorem 1.6.—A goal which we achieve, at least partially, in Chapter 5. However, the intermediate results we need might be of sufficient interest in their own right, as to warrant our devoting a separate chapter to them.

Most of what we do here was worked out by Baier [4] for the case of Gaussian integers $\mathcal{O} = \mathbb{Z}[i]$ with generator $\omega = i$ and our proofs follow his arguments quite closely.

4.1 Introduction

4.1.1 Approximation with rational primes

Before discussing Diophantine approximation with primes from $\mathcal{O}$, we describe the origin of the problem. Given some irrational ϑ, Dirichlet's approximation theorem asserts the existence of infinitely many integers a, q ($q \neq 0$) with

$$\left| \vartheta - \frac{a}{q} \right| < \frac{1}{q^2},$$

or, equivalently, on writing $\|\rho\| = \min_{x \in \mathbb{Z}} |\rho - x|$ for the distance to a nearest integer,

$$\|q\vartheta\| < q^{-1}. \tag{4.1}$$

On the other hand, Hurwitz's approximation theorem implies that the exponent -1 in (4.1) is optimal in the sense that it cannot be decreased without the resulting new inequality failing to admit infinitely many solutions for some irrational ϑ (see, e.g., [31, Theorems 193 and 194]).—In this regard the reader may also check the discussion at the end of Section 2.3.

A natural variation on the question about the solubility of (4.1) is to impose the additional restriction that q be prime and ask for which exponent θ one is able to establish that, for any irrational ϑ,

$$\|p\vartheta\| < p^{-\theta} \quad \text{for infinitely many primes } p. \tag{4.2}$$

In this direction Vinogradov [89] obtained (4.2) with $\theta = \frac{1}{5} - \epsilon$, a result which has since then been improved by a number of researchers (see Table 4.1) culminating in the work of Matomäki [63] who obtained $\theta = 1/3 - \epsilon$. This exponent is considered to be the limit of the current technology (see the comments in [40]).

Date	Author(s)	θ	
1978	Vaughan [86]	$1/4 - \epsilon$	$= 0.25 - \epsilon$
1983	Harman [33]	$3/10$	$= 0.3$
1993	Jia [47]	$4/13$	$= 0.3076\ldots$
1996	Harman [34]	$7/22$	$= 0.31\overline{81}$
2000	Jia [48]	$9/28$	$= 0.3214\ldots$
2002	Heath-Brown and Jia [40]	$16/49$	$= 0.3265\ldots$
2009	Matomäki [63]	$1/3 - \epsilon$	$= 0.3\overline{3} - \epsilon$

Table 4.1: Improvements on the admissible exponent θ in (4.2).

4.1.2 The Gaussian case

In view of the above, Baier [4] proposed to study the analogue of (4.2) in the setting of Gaussian integers.

Before stating his results we need to fix some notation which will be used throughout. Recall that $\mathbb{K} \subset \mathbb{C}$ is an imaginary quadratic number field embedded into $\mathbb{C}$ and $\mathcal{O}$ denotes its ring of integers. It is a classical result from algebraic number theory that $\mathcal{O}$ is a free $\mathbb{Z}$-module of rank $\dim_{\mathbb{Q}} \mathbb{K} = 2$. Thus, there is some ω such that $\{1, \omega\}$ is a $\mathbb{Z}$-basis of $\mathcal{O}$. Since, by assumption, $\mathbb{K} \not\subseteq \mathbb{R}$, and $\mathbb{K}$ being the field of fractions of $\mathcal{O}$, it follows that $\Im\omega \neq 0$. In particular, $\{1, \omega\}$ turns out to be an $\mathbb{R}$-basis of $\mathbb{C}$ and, given some $\vartheta \in \mathbb{C}$, we write $\Re_\omega\vartheta$ and $\Im_\omega\vartheta$ for the unique real numbers satisfying $\vartheta = \Re_\omega\vartheta + (\Im_\omega\vartheta)\omega$. With this notation, we put

$$\|\vartheta\|_\omega = \max\{\|\Re_\omega\vartheta\|, \|\Im_\omega\vartheta\|\}.$$

The *norm* $N(m)$ of an element $m \in \mathcal{O}$ is defined to be the number of elements in the factor ring $\mathcal{O}/m\mathcal{O}$. One can show that

$$N(m) = |m|^2.$$

Some further basic properties of $\mathcal{O}$ are recorded in Appendix A and we freely refer to them throughout.

We now state Baier's results:

Theorem 4.1 (Baier [4]). *Let ϑ be a complex number such that $\vartheta \notin \mathbb{Q}(i)$ and $\epsilon > 0$ be an arbitrary constant. Then there exists an infinite increasing sequence of natural numbers $(x_k)_{k\in\mathbb{N}}$ such that*

$$\sum_{\substack{x_k/2 \leq N(p) < x_k \\ \|p\vartheta\|_i \leq \delta_k}} 1 \sim 4\delta_k^2 \cdot \sum_{x_k/2 \leq N(p) < x_k} 1 \quad \text{as } k \to \infty$$

if $x_k^{-1/24+\epsilon} \leq \delta_k \leq 1/2$; here the summation is over Gaussian primes p.

Corollary 4.2 (Baier [4]). *On the hypotheses of Theorem 4.1, there exist infinitely many Gaussian primes p such that*

$$\|p\vartheta\|_i \leq N(p)^{-1/24+\epsilon}.$$

Remark 4.3. *There appear to be some inaccuracies in [4]. As a consequence, the exponent $1/24$ in Theorem 4.1 and Corollary 4.2 should be halved, giving $1/48$. The precise reason for this is elaborated in Remark 4.18 below.*

4.2 Main results: from $\mathbb{Z}[i]$ to general $\mathcal{O}$

The problem discussed in Section 4.1.1 is not quite sufficient for finding primes in all ordinary Beatty sets $\mathcal{B}(\alpha, \beta)$ ($\beta \geq 0$) due to the interval (mod 1) in Lemma 1.2 not being centred around 0. This requires one to replace the inequality in (4.2) with the following shifted version:

$$\|p\vartheta + \tilde{\beta}\| < p^{-\theta} \quad \text{(with } \vartheta = \alpha^{-1} \text{ and } \tilde{\beta} \in \mathbb{R}\text{)}. \tag{4.3}$$

In fact, most authors [86, 33, 47, 34, 48] listed in Table 4.1 considered this more general problem, but the innovation introduced in the work of Heath-Brown and Jia [40] has, as they remark, the "defect" of entailing the restriction to $\tilde{\beta} = 0$. In turn, Matomäki [63] also only treats the case $\tilde{\beta} = 0$.

For our application in Chapter 5 we also need the shifted version in the setting of $\mathcal{O}$. A trivial modification in the proof given by Baier [4] allows one to include the shift $\tilde{\beta}$ and, for the most part, the modifications necessary to make the proof work for general $\mathcal{O}$ with generator ω are straightforward. However, we note that there are some instances where one easily runs into problems when not working with the "correct" parametrisation of $\mathcal{O}$; we invite the reader to keep this comment in mind when looking at Remark 4.10 and Section 4.5.4.

Theorem 4.4. *Suppose that $\mathbb{K}$ has class number 1. Let ϑ be a complex number not contained in $\mathbb{K}$. Furthermore, suppose that one has coprime $a, q \in \mathcal{O}$ such that*

$$q \neq 0, \quad \frac{a}{q} \notin \mathcal{O}, \quad \text{and} \quad \gamma = \vartheta - \frac{a}{q} \quad \text{satisfies} \quad |\gamma| \leq \frac{C}{|q|^2} \tag{4.4}$$

for some constant $C > 0$. Put $x = |q|^{12}$ and let $\epsilon > 0$ be an arbitrary constant. Then, for any complex number β and any δ such that

$$x^{-1/48+4\epsilon} \leq \delta \leq \tfrac{1}{2}, \tag{4.5}$$

the difference

$$\sum_{\substack{x/2 \leq N(p) < x \\ \|p\vartheta + \tilde{\beta}\|_\omega \leq \delta}} 1 - 4\delta^2 \sum_{x/2 \leq N(p) < x} 1$$

does not exceed

$$2^{65} C^2 |\omega|^{14} \epsilon^{-8} M_{\mathcal{O}}(\epsilon)^{5/2} \delta^2 x^{1-\epsilon}$$

in absolute value. Here $M_{\mathcal{O}}(\epsilon)$ is given as in Lemma A.8 below.

To establish the existence of a and q as required by the theorem, one can use a result due to Hilde Gintner [28]. In this regard, let Λ be the *fundamental parallelogram* spanned by 1 and ω,

$$\Lambda = \{\lambda_1 + \lambda_2\omega : \lambda_1, \lambda_2 \in [0,1)\}. \tag{4.6}$$

Theorem 4.5 (Gintner [28]). *Let ϑ be a complex number not contained in $\mathbb{K}$. Then there are infinitely many $a, q \in \mathcal{O}$ satisfying (4.4) with*

$$C = \frac{\sqrt{6}}{\pi} \operatorname{area} \Lambda$$

and Λ given by (4.6).

The above theorem does not assert that a, q be coprime. However, if $\mathbb{K}$ has class number 1, then one can appeal to unique factorisation and cancel any potential non-trivial common factors from a and q. Upon using Corollary A.6, we immediately obtain the following:

Corollary 4.6. *On implementing the changes enunciated by Remark 4.3, the conclusions of Theorem 4.1 and Corollary 4.2 remain true if one replaces*

- *$\mathbb{Q}(i)$ with some $\mathbb{K}$ with class number 1,*

- *"Gaussian primes" with "primes from $\mathcal{O}$."*

4.3 Plan of the proof

The proof of Theorem 4.4 rests upon a sieve result due to Harman in the setting of $\mathcal{O}$. The sieve method itself is also capable of yielding lower bounds instead of asymptotic formulae in exchange for the prospect of increasing the admissible range for δ in (4.5), but we does not implement this here. The following special case suffices for our purposes:

Theorem 4.7 (Special case of Harman's sieve for $\mathcal{O}$). *Suppose that $\mathcal{O}$ has class number 1. Let $x \geq 3$ be real and suppose that $\mathscr{A}$ is some subset of*

$$\mathscr{B} = \{m \in \mathcal{O} : x/2 \leq N(m) < x\}. \tag{4.7}$$

Suppose further that one has numbers $\lambda, Y_I, Y_{II} > 0$, $0 < \mu \leq 1$, $0 < \kappa \leq \frac{1}{2}$, and $M > x^\mu$ with the following property:
For any sequences $(a_m)_{m \in \mathcal{O}}$, $(b_n)_{n \in \mathcal{O}}$ of complex numbers with $|a_m| \leq 1$ and $|b_n| \leq d((n))$ with $d(\ldots)$ as defined in Appendix A.4, one has

$$\left| \sum_{\substack{mn \in \mathscr{A} \\ N(m) < M}} \sum a_m - \lambda \sum_{\substack{mn \in \mathscr{B} \\ N(m) < M}} \sum a_m \right| \leq Y_I, \tag{4.8}$$

$$\left| \sum_{\substack{mn \in \mathscr{A} \\ x^\mu < N(m) < x^{\mu+\kappa}}} \sum a_m b_n - \lambda \sum_{\substack{mn \in \mathscr{B} \\ x^\mu < N(m) < x^{\mu+\kappa}}} \sum a_m b_n \right| \leq Y_{II}. \tag{4.9}$$

Then

$$|S(\mathscr{A}, x^{\kappa}) - \lambda S(\mathscr{B}, x^{\kappa})| \leq Y_{\mathrm{I}} + 2^{11}(Y_{\mathrm{II}} + 2)(\log x)^3,$$

where

$$S(\mathscr{A}, z) = \#\{n \in \mathscr{A} : p \nmid n \text{ for all primes } p \in \mathcal{O} \text{ with } N(p) < z\}.$$

The proof of Theorem 4.7 is given in Appendix B.

Remark 4.8. *The reader will certainly observe that when $\kappa = \frac{1}{2}$, Theorem 4.7 is detecting primes:*

$$S(\mathscr{A}, \sqrt{x}) = \#\{p \in \mathscr{A} : p \text{ prime}\}.$$

However, the above set may contain associates.

Returning to the goal of proving Theorem 4.4, we shall verify the hypotheses of Theorem 4.7 with $\kappa = \frac{1}{2}$. This is achieved in Section 4.5, but requires some exponential sum estimates which we shall derive first.

4.4 *Prélude:* exponential sum estimates

In Section 4.5 we need estimates for sums of the shape

$$\sum_n \sum_m \mathrm{e}(\Im_{\omega}(mn\vartheta)),$$

where the summation over m is restricted to some annulus $x(n) \leq N(m) \leq y(n)$ with bounds $x(n)$ und $y(n)$ depending on n. In Section 4.4.1 we estimate the inner summation and in Section 4.4.2 we deal with the additional summation over n. The corresponding proof is then carried out in the subsequent sections.

We add that our bias towards working with $\Im_{\omega}$ instead of $\Re_{\omega}$ will be justified in Remark 4.19 below. Moreover, in what follows, we sometimes have expressions like $1/\|\Re_{\omega}\vartheta\|$; if a division by zero occurs there, then the result is understood to mean $+\infty$.

4.4.1 Basic estimates for linear exponential sums

Lemma 4.9. *Let ϑ be a complex number and suppose that one has numbers x, y such that $0 \leq x \leq y$. Then*

$$\left| \sum_{x \leq N(m) \leq y} \mathrm{e}(\Im_{\omega}(m\vartheta)) \right| \leq 13|\omega|^2 \sqrt{y} \min\left\{ \sqrt{y}, \frac{1}{\|\Re_{\omega}\vartheta\|}, \frac{1}{\|\Im_{\omega}\vartheta\|} \right\}. \tag{4.10}$$

Proof. We may assume $y \geq 1$, for (4.10) is trivial otherwise. We denote the sum on the left hand side of (4.10) by $S(x, y)$. On writing

$$\omega^2 = \xi_1 + \xi_2\omega \quad (\xi_1 = \Re_{\omega}\omega^2, \quad \xi_2 = \Im_{\omega}\omega^2), \tag{4.11}$$

37

$m = m_1 + m_2\omega$ and $\ell_1 = m_1 + m_2\xi_2$, we obtain the following two expressions for $\Im_\omega(m\vartheta)$:

$$\Im_\omega(m\vartheta) = m_2(\Re_\omega\vartheta + \xi_2\Im_\omega\vartheta) + m_1\Im_\omega\vartheta$$
$$= \ell_1\Im_\omega\vartheta + m_2\Re_\omega\vartheta.$$

Therefore,

$$S(0,y) = \begin{cases} \displaystyle\sum_{m_1}\sum_{\substack{m_2 \\ 0 \le N(m_1+m_2\omega) \le y}} e(m_2(\Re_\omega\vartheta + \xi_2\Im_\omega\vartheta))\, e(m_1\Im_\omega\vartheta), \\[2em] \displaystyle\sum_{\ell_1}\sum_{\substack{m_2 \\ 0 \le N(\ell_1-m_2\xi_2+m_2\omega) \le y}} e(\ell_1\Im_\omega\vartheta)\, e(m_2\Re_\omega\vartheta). \end{cases} \tag{4.12}$$

So, on recalling the well-known bound

$$\left|\sum_{a \le j \le b} e(j\rho)\right| \le \frac{1}{2\|\rho\|} \quad (a,b,\rho \in \mathbb{R}),$$

and using the triangle inequality on the outer summations in (4.12),

$$|S(0,y)| \le \frac{\#\{m_2 : \exists m_1 \text{ s.t. } 0 \le N(m_1 + m_2\omega) \le y\}}{2\|\Im_\omega\vartheta\|}, \tag{4.13}$$

$$|S(0,y)| \le \frac{\#\{\ell_1 : \exists m_2 \text{ s.t. } 0 \le N(\ell_1 - m_2\xi_2 + m_2\omega) \le y\}}{2\|\Re_\omega\vartheta\|}. \tag{4.14}$$

The numerators here are bounded easily: indeed, since

$$N(m_1 + m_2\omega) \ge m_2^2(\Im\omega)^2,$$

using $y \ge 1$ and (A.4), one easily bounds the numerator in (4.13) by $4\sqrt{y}$. On the other hand for $|\ell_1| > c\sqrt{y}$ with $c = (2 + \frac{2}{\sqrt{3}})|\omega|^2$ there is no m_2 such that

$$y \ge N(\ell_1 - m_2\xi_2 + m_2\omega)$$
$$\ge \max\{|\ell_1 - m_2\xi_2 + m_2\Re\omega|^2, m_2^2(\Im\omega)^2\}, \tag{4.15}$$

for otherwise it would follow that

$$\sqrt{y} > c\sqrt{y} - |m_2| \cdot |\xi_2 - \Re\omega|$$

so that $|m_2| \cdot |\xi_2 - \Re\omega| > (c-1)\sqrt{y}$, but then

$$|m_2\Im\omega| > \frac{c-1}{|\xi_2| + |\Re\omega|}\sqrt{y} \ge \frac{(1 + 2/\sqrt{3})|\omega|^2}{|\omega|^2 + |\omega|}\sqrt{y}$$
$$= \frac{1 + 2/\sqrt{3}}{1 + |\omega|^{-1}}\sqrt{y} \ge \sqrt{y},$$

in contradiction to (4.15). Hence, the numerator in (4.14) is bounded by $2c\sqrt{y}+1 \leq 8|\omega|^2\sqrt{y}$. Thus,

$$|S(0,y)| \leq 8|\omega|^2\sqrt{y}\min\left\{\frac{1}{2\|\Im_\omega\vartheta\|}, \frac{1}{2\|\Re_\omega\vartheta\|}\right\}$$

and, together with the trivial bound $|S(x,y)| \leq 13y$ from Lemma A.3, this is clearly satisfactory to establish (4.10) for $x = 0$, and the case $x > 0$ follows from this bound and

$$S(x,y) = S(0,y) - \lim_{\epsilon\searrow0} S(0,x-\epsilon). \qquad \square$$

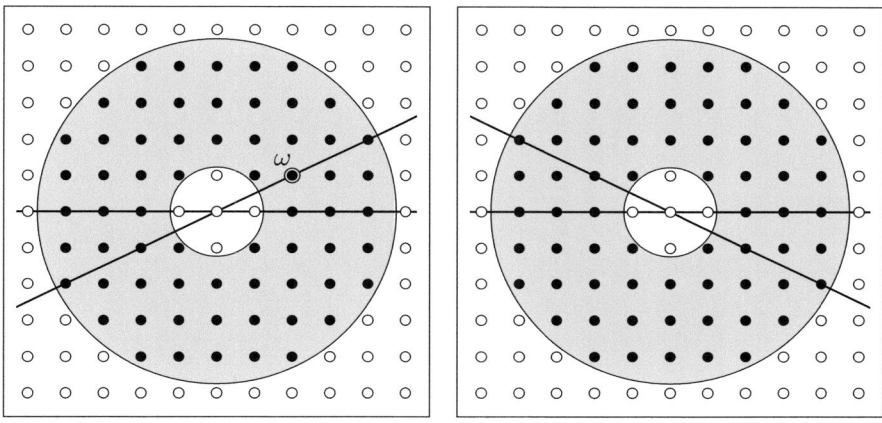

Figure 6: Illustration of the summation over m in (4.10) with $\mathbb{K} = \mathbb{Q}(\omega)$ and $\omega = 2+i$. One is summing over points in an annulus. The inner summations in (4.12) are over the points on lines (and their translates) of the type drawn in the figure on the right. The figure on the left shows the lines corresponding to summing over m_1, m_2 ($m_1 + m_2\omega$).

Remark 4.10. *The reader will note that the summation along ℓ_1 in the above proof (see Fig. 6) is not a mere cosmetic tool, but actually has a qualitative effect: taking the perhaps more natural approach of summing over m_2 and with m_1 fixed (and the other way around) only supplies the bound*

$$\left|\sum_{x\leq N(m)\leq y} e(\Im_\omega(m\vartheta))\right| \ll_\omega \sqrt{y}\min\left\{\sqrt{y}, \frac{1}{\|\Im_\omega\vartheta\|}, \frac{1}{\|\Im_\omega(\omega\vartheta)\|}\right\},$$

which is a dead-end, since for this an analogue of Lemma 4.13 below cannot be obtained.

4.4.2 Distribution of fractional parts

In view of Lemma 4.9 and recalling the goal stated at the beginning of Section 4.4, we are faced with the problem of estimating sums of the shape

$$\sum_{n\in\mathscr{X}} E(n,M), \qquad (4.16)$$

39

where $M \geq 1$, $\mathscr{X}$ is some subset of $\{n \in \mathcal{O} : 1 \leq N(n) < x\}$ with $x \geq 1$, and

$$E(n, M) = \min\left\{ M, \frac{1}{\|\Re_\omega(n\vartheta)\|}, \frac{1}{\|\Im_\omega(n\vartheta)\|} \right\}. \qquad (4.17)$$

The usual attack against such a problem is to replace ϑ by some Diophantine approximation a/q. Subsequently, after bounding the error introduced from the approximation, one is able to control averages of $E(n, M)$ with n constrained to boxes (say) not too large in terms of $|q|$. By splitting the full range of n in (4.16) into such boxes, one derives a bound for (4.16) of the shape seen in Theorem 4.11 below.

The theorem we shall prove may be stated as follows:

Theorem 4.11. *Let $\mathscr{X}$ be a subset of all $n \in \mathcal{O}$ with $1 \leq N(n) < x$ and suppose that one has coprime $a, q \in \mathcal{O}$ satisfying (4.4). Put*

$$S = \sum_{n \in \mathscr{X}} \min\left\{ M, \frac{1}{\|\Re_\omega(n\vartheta)\|}, \frac{1}{\|\Im_\omega(n\vartheta)\|} \right\}.$$

Then, assuming $M \geq 1$,

$$S \leq 2^{20}(1 + C^2|\omega|^4 x|q|^{-2})(M + |\omega|^2|q|^2 \log(2M)). \qquad (4.18)$$

Furthermore, if $x \leq |q|^2/(12C|\omega|^2)^2$, then

$$S \leq 2^{15}|\omega|^2|q|^2 \log(4|\omega||q|). \qquad (4.19)$$

The proof of this result follows the outline given above and is undertaken in the next two subsections.

Remark 4.12. *Our bounds (4.18), (4.19) are slightly sharper than the corresponding ones given in [4, (39) and (40)]: the power of our logarithmic factor is 1 instead of 2, and no term $|q|\sqrt{M}$ appears in (4.19). We shall comment on these improvements below in Remark 4.15.*

4.4.3 Diophantine lemmas

As a first step, we show that

$$\max\{\|\Re_\omega\vartheta\|, \|\Im_\omega\vartheta\|\} \qquad (4.20)$$

cannot be too small if $\vartheta \in \mathbb{K}$ is not an algebraic integer. At this point the reader may wish to reflect on Remark 4.10 once more by convincing him- or herself that what follows would not work if $\Re_\omega\vartheta$ in (4.20) were to be replaced by $\Im_\omega(\omega\vartheta)$.

Lemma 4.13. *For non-zero $a, q \in \mathcal{O}$ such that $\vartheta = a/q \notin \mathcal{O}$, it holds that*

$$\max\{\|\Re_\omega\vartheta\|, \|\Im_\omega\vartheta\|\} \geq \frac{1}{2|\omega||q|}.$$

Proof. Pick $m = m_1 + m_2\omega \in \mathcal{O}$ such that

$$\|\mathfrak{R}_\omega\vartheta\| = |\mathfrak{R}_\omega\vartheta - m_1| \quad \text{and} \quad \|\mathfrak{I}_\omega\vartheta\| = |\mathfrak{I}_\omega\vartheta - m_2|.$$

Now certainly it holds that

$$\begin{aligned}
|\vartheta - m| &= |(\mathfrak{R}_\omega\vartheta - m_1) + (\mathfrak{I}_\omega\vartheta - m_2)\omega| \\
&\leq 2|\omega|\max\{|\mathfrak{R}_\omega\vartheta - m_1|, |\mathfrak{I}_\omega\vartheta - m_2|\} \\
&= 2|\omega|\max\{\|\mathfrak{R}_\omega\vartheta\|, \|\mathfrak{I}_\omega\vartheta\|\}.
\end{aligned}$$

Therefore, to prove the lemma, it suffices to give a suitable lower bound for $|\vartheta - m|$, which, upon noting that $\vartheta \notin \mathcal{O}$, is quite easy:

$$|\vartheta - m| = |q|^{-1}|a - qm| \geq |q|^{-1}\min_{0 \neq r \in \mathcal{O}}|r| = |q|^{-1}. \qquad \square$$

Next, we intend to derive a result similar to Lemma 4.13, when ϑ is slightly perturbed:

Lemma 4.14. *Let ϑ be a complex number and a, q be such that (4.4) holds. Furthermore, suppose that $n \in \mathcal{O}$ satisfies*

$$|n| \leq \frac{|q|}{12C|\omega|^2}$$

and na be indivisible by q. Then

$$\max\{\|\mathfrak{R}_\omega(n\vartheta)\|, \|\mathfrak{I}_\omega(n\vartheta)\|\} \geq \frac{1}{4|\omega||q|}.$$

Proof. First, we separate the perturbation γ from the rest: we have

$$\begin{aligned}
\|\mathfrak{I}_\omega(n\vartheta)\| &= \min_{k \in \mathbb{Z}}|\mathfrak{I}_\omega(na/q) - k + \mathfrak{I}_\omega(n\gamma)| \\
&\geq \min_{k \in \mathbb{Z}}|\mathfrak{I}_\omega(na/q) - k| - |\mathfrak{I}_\omega(n\gamma)| \\
&= \|\mathfrak{I}_\omega(na/q)\| - |\mathfrak{I}_\omega(n\gamma)|,
\end{aligned}$$

and the same holds when one replaces $\mathfrak{I}_\omega$ by $\mathfrak{R}_\omega$. The last term therein is bounded easily: using $|\xi| \geq |\mathfrak{I}_\omega\xi| \cdot |\mathfrak{I}\omega|$, (A.4) and writing $N = C|n|/|q|^2$ for the moment, we have

$$|\mathfrak{I}_\omega(n\gamma)| \leq \max_{\substack{\xi \in \mathbb{C} \\ |\xi| \leq N}}|\mathfrak{I}_\omega\xi| \leq \frac{N}{|\mathfrak{I}\omega|} \leq \frac{2N}{\sqrt{3}}.$$

A similar calculation also bounds the corresponding $\mathfrak{R}_\omega$-term:

$$\begin{aligned}
|\mathfrak{R}_\omega(n\gamma)| &\leq \max\{|\rho_1| : \rho_1, \rho_2 \in \mathbb{R}, \sqrt{(\rho_1 + \rho_2\mathfrak{R}\omega)^2 + (\rho_2\mathfrak{I}\omega)^2} \leq N\} \\
&\leq \max\{|\theta| + |\rho_2\mathfrak{R}\omega| : \theta, \rho_2 \in \mathbb{R}, \sqrt{\theta^2 + (\rho_2\mathfrak{I}\omega)^2} \leq N\} \\
&\leq (1 + |\mathfrak{R}\omega/\mathfrak{I}\omega|)N \leq (1 + 2/\sqrt{3})|\omega|N.
\end{aligned}$$

41

Thus, using Lemma 4.13,

$$\max\{\|\Re_\omega(n\vartheta)\|, \|\Im_\omega(n\vartheta)\|\}$$
$$\geq \max\{\|\Re_\omega(na/q)\|, \|\Im_\omega(na/q)\|\} - 3|\omega|N$$
$$\geq \frac{1}{2|\omega||q|} - 3|\omega|N = \frac{1}{|q|}\left(\frac{1}{2|\omega|} - 3C|\omega|\frac{|n|}{|q|}\right).$$

Now, by assumption, the term in the parentheses is $\geq (4|\omega|)^{-1}$, and the assertion of the lemma follows. $\qquad\square$

4.4.4 Proof of Theorem 4.11

Assume the hypotheses of Theorem 4.11 and let n and $\tilde{n}$ be two distinct algebraic integers in $\mathcal{O}$ which coincide modulo q. Then there is some non-zero $m \in \mathcal{O}$ such that $n - \tilde{n} = mq$ and, hence, $|n - \tilde{n}| = |m||q| \geq |q|$. Assuming a and q to be coprime and $|n - \tilde{n}| < |q|$, we conclude that $(n - \tilde{n})a$ is divisible by q if and only if $n = \tilde{n}$. Consequently, if $\mathscr{R} \subseteq \mathbb{C}$ is some set with

$$\operatorname{diam}\mathscr{R} \leq \frac{|q|}{12C|\omega|^2}, \tag{4.21}$$

then, according to Lemma 4.14, any two distinct points $n\vartheta, \tilde{n}\vartheta$ ($n, \tilde{n} \in \mathscr{X} \cap \mathscr{R}$) satisfy the spacing condition

$$\max\{\|\Re_\omega((n - \tilde{n})\vartheta)\|, \|\Im_\omega((n - \tilde{n})\vartheta)\|\} \geq \frac{1}{4|\omega||q|}.$$

Therefore, for $0 < \Delta_1, \Delta_2 \leq \frac{1}{2}$, the sum

$$\sum_{\substack{n\in\mathscr{X}\cap\mathscr{R} \\ \|\Re_\omega(n\vartheta)\|\leq\Delta_1 \\ \|\Im_\omega(n\vartheta)\|\leq\Delta_2}} 1 = \sum_{\substack{n\in\mathscr{X}\cap\mathscr{R} \\ \{\Re_\omega(n\vartheta)\}\leq\Delta_1 \\ \{\Im_\omega(n\vartheta)\}\leq\Delta_2}} 1 + \sum_{\substack{n\in\mathscr{X}\cap\mathscr{R} \\ \{\Re_\omega(n\vartheta)\}\geq1-\Delta_1 \\ \{\Im_\omega(n\vartheta)\}\leq\Delta_2}} 1 + \sum_{\substack{n\in\mathscr{X}\cap\mathscr{R} \\ \{\Re_\omega(n\vartheta)\}\leq\Delta_1 \\ \{\Im_\omega(n\vartheta)\}\geq1-\Delta_2}} 1 + \sum_{\substack{n\in\mathscr{X}\cap\mathscr{R} \\ \{\Re_\omega(n\vartheta)\}\geq1-\Delta_1 \\ \{\Im_\omega(n\vartheta)\}\geq1-\Delta_2}} 1$$

is bounded by four times the maximum number of points of pairwise maximum norm distance $\geq (4|\omega||q|)^{-1}$ that can be put in a rectangle with side lengths Δ_1 and Δ_2, i.e.,

$$\sum_{\substack{n\in\mathscr{X}\cap\mathscr{R} \\ \|\Re_\omega(n\vartheta)\|\leq\Delta_1 \\ \|\Im_\omega(n\vartheta)\|\leq\Delta_2}} 1 \leq 4(1 + 4|\omega||q|\Delta_1)(1 + 4|\omega||q|\Delta_2). \tag{4.22}$$

Moving on, let $L \in \mathbb{N}$ be a parameter at our disposal. Then the sum

$$S(\mathscr{R}) = \sum_{n\in\mathscr{X}\cap\mathscr{R}} E(n, M)$$

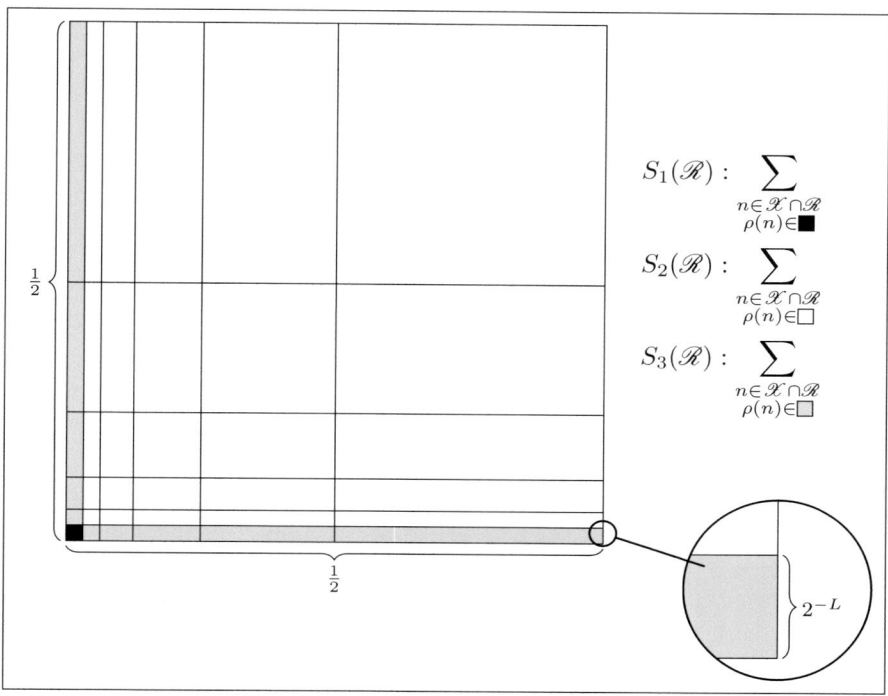

$$S_1(\mathscr{R}) : \sum_{\substack{n \in \mathscr{X} \cap \mathscr{R} \\ \rho(n) \in \blacksquare}}$$

$$S_2(\mathscr{R}) : \sum_{\substack{n \in \mathscr{X} \cap \mathscr{R} \\ \rho(n) \in \square}}$$

$$S_3(\mathscr{R}) : \sum_{\substack{n \in \mathscr{X} \cap \mathscr{R} \\ \rho(n) \in \square}}$$

Figure 7: The grouping of $n \in \mathscr{X} \cap \mathscr{R}$ from the summation in $S(\mathscr{R})$ into $S_1(\mathscr{R})$, $S_2(\mathscr{R})$ or $S_3(\mathscr{R})$ depends on which region the point $\rho(n) = (\|\Re_\omega(n\vartheta)\|, \|\Im_\omega(n\vartheta)\|)$ belongs to: the black square, a white rectangle or a grey rectangle.

admits a decomposition (compare Fig. 7)

$$S(\mathscr{R}) \leq \sum_{\substack{n \in \mathscr{X} \cap \mathscr{R} \\ \|\Re_\omega(n\vartheta)\| \leq 2^{-L} \\ \|\Im_\omega(n\vartheta)\| \leq 2^{-L}}} M + \sum_{\substack{2 \leq k_1, k_2 \leq L, \, n \in \mathscr{X} \cap \mathscr{R} \\ 2^{-k_1} < \|\Re_\omega(n\vartheta)\| \leq 2^{1-k_1} \\ 2^{-k_2} < \|\Im_\omega(n\vartheta)\| \leq 2^{1-k_2}}} \min\{2^{k_1}, 2^{k_2}\} +$$

$$+ \sum_{2 \leq k \leq L} \left\{ \sum_{\substack{n \in \mathscr{X} \cap \mathscr{R} \\ 2^{-k} < \|\Re_\omega(n\vartheta)\| \leq 2^{1-k} \\ \|\Im_\omega(n\vartheta)\| \leq 2^{-L}}} + \sum_{\substack{n \in \mathscr{X} \cap \mathscr{R} \\ \|\Re_\omega(n\vartheta)\| \leq 2^{-L} \\ 2^{-k} < \|\Im_\omega(n\vartheta)\| \leq 2^{1-k}}} \right\} 2^k$$

$$= S_1(\mathscr{R}) + S_2(\mathscr{R}) + S_3(\mathscr{R}), \quad \text{say.}$$

By (4.22) and using $(a+b)^2 \leq 2a^2 + 2b^2$ $(a, b \geq 1)$,

$$S_1(\mathscr{R}) \leq 4(1 + 4|\omega||q|2^{-L})^2 M$$
$$\leq 8M + 32|\omega|^2|q|^2 2^{-2L} M.$$

Moreover, using

$$\min\{2^{k_1}, 2^{k_2}\} \leq \sqrt{2^{k_1 + k_2}} \qquad (4.23)$$

43

and (4.22),

$$S_2(\mathcal{R}) \leq \left(\sum_{2 \leq k \leq L} 2^{k/2} \cdot 2(1 + 4|\omega||q|2^{-k}) \right)^2$$
$$\leq 94 \cdot 2^L + 24|\omega|^2|q|^2. \tag{4.24}$$

Similarly,

$$S_3(\mathcal{R}) \leq 8 \sum_{2 \leq k \leq L} 2^k (1 + 4|\omega||q|2^{-k})(1 + 4|\omega||q|2^{-L})$$
$$\leq 16 \cdot 2^L + 128|\omega|^2|q|^2 L.$$

Assuming $M \geq 1$, we take $L = \lceil \frac{1}{2\log 2} \log(2M) \rceil$ to obtain

$$S(\mathcal{R}) \leq 355(M + |\omega|^2|q|^2 \log(2M)). \tag{4.25}$$

Additionally, if
$$x \leq |q|^2/(12C|\omega|^2)^2,$$

then we take $L = \lceil \frac{1}{\log 2} \log(4|\omega||q|) \rceil$. In this case Lemma 4.14 shows that $S_1(\mathcal{R})$ vanishes and, consequently, we have

$$S(\mathcal{R}) \leq 930|\omega|^2|q|^2 \log(4|\omega||q|). \tag{4.26}$$

Finally, we note that the set $\mathcal{X}$ can be covered by fewer than

$$\left(1 + \frac{2\sqrt{x}}{(\text{diam. bound})/\sqrt{2}}\right)^2 \leq 2 + 2304\, C^2|\omega|^4 x|q|^{-2}$$

squares $\mathcal{R}$ with diameter (4.21). Together with (4.25) this proves (4.18), and together with (4.26) we obtain (4.19). This proves the theorem.

Remark 4.15. *Continuing Remark 4.12, we note that the aforementioned improvements have the following two origins respectively: First, the estimation of $S_2(\mathcal{R})$ in (4.24) does not require a factor L^2. Second, using (4.23) when estimating $S_3(\mathcal{R})$ is more wasteful than necessary. In any case, however, the overall quality of the final result is not affected in the sense that the exponent $\frac{1}{48}$ in (4.5) can be obtained even without the aforementioned improvement.*

4.5 The input for Harman's sieve

Here we continue to adapt the work of Baier [4] and work on verifying the hypotheses of Theorem 4.7 in our setting.

4.5.1 Setting up linear and bilinear forms

Let $\mathscr{B}$ be given as in (4.7) with $x \geq 3$. We shall want to estimate sums of the type

$$\sum_{mn \in \mathscr{A}} \sum a_m b_n, \tag{4.27}$$

where

- $\mathscr{A} = \{n \in \mathscr{B} : \|n\vartheta + \tilde{\beta}\|_\omega \leq \delta\}$,

- the summation indices m, n vary through $\mathcal{O}$,

- the coefficient sequences $(a_m)_m$ and $(b_n)_n$ consist of complex numbers and satisfy $|a_n| \leq 1$ and $|b_n| \leq d((n))$.

To be more specific, for parameters $\mu > 0$ and $0 < \kappa \leq \frac{1}{2}$, there are two types of sums we would like to estimate

- *Type I*: b_n takes only either of the values 0 or 1 and $(a_m)_m$ is supported only on m with $N(m) < M$ for some M with $x^\mu < M \leq x$ (see (4.8)).

- *Type II*: $(a_m)_m$ is supported only on m with $x^\mu \leq N(m) < x^{\mu+\kappa}$ (see (4.9)).

Each type requires a different treatment, but for now it is convenient to start by transforming (4.27) without restricting to either of the above types. We start with the following result which furnishes a good finite Fourier approximation to the *saw-tooth function* ψ given by

$$\psi(t) = t - \lfloor t \rfloor - \tfrac{1}{2} \quad (t \in \mathbb{R}). \tag{4.28}$$

Lemma 4.16 (Vaaler [84]). *For $0 < |\rho| < 1$ let*

$$W(\rho) = \pi\rho(1 - |\rho|)\cot(\pi\rho) + |\rho|.$$

Let J be a positive integer and t a real number. Write

$$\psi^*(t) = -\sum_{1 \leq |j| < J} \frac{W(j/J)}{2\pi i j}\, \mathrm{e}(jt)$$

and

$$\sigma(t) = \frac{1}{2J} + \frac{1}{2J}\sum_{1 \leq |j| < J} (1 - |j|/J)\,\mathrm{e}(jt).$$

Then σ is real-valued, non-negative, and we have

$$|\psi^*(t) - \psi(t)| \leq \sigma(t)$$

for all real numbers t, where $\mathrm{e}(t) = \exp(2\pi i t)$ and ψ is given by (4.28).

Proof. See [84, Theorem 18] or, e.g., the appendix in [29]. $\qquad\square$

The particular shape of W from Lemma 4.16 does not concern us. Indeed, the following trivial bound suffices for our purposes:

Lemma 4.17. *For $0 < |t| < 1$, $0 \le W(t) \le 1$.*

We now derive a useful expansion of the characteristic function $\mathbf{1}_{\mathscr{A}}$ of $\mathscr{A}$ evaluated at algebraic integers. Indeed, letting y denote any element of $\mathcal{O}$ and writing

$$\tilde{\beta} = \tilde{\beta}_1 + \tilde{\beta}_2 \omega, \quad x_{1,y} = \Re_\omega(y\vartheta) \quad \text{and} \quad x_{2,y} = \Im_\omega(y\vartheta),$$

we have

$$\begin{aligned}
\mathbf{1}_{\mathscr{A}}(y) &= \prod_{k=1,2} \left(2\delta + (\psi(\tilde{\beta}_k - \delta - x_{k,y}) - \psi(\tilde{\beta}_k + \delta - x_{k,y}))\right) \\
&= 4\delta^2 + 2\delta \sum_{k=1,2} (\psi(\tilde{\beta}_k - \delta - x_{k,y}) - \psi(\tilde{\beta}_k + \delta - x_{k,y})) + \\
&\quad + \prod_{k=1,2} (\psi(\tilde{\beta}_k - \delta - x_{k,y}) - \psi(\tilde{\beta}_k + \delta - x_{k,y})) \\
&= 4\delta^2 + 2\delta \sum_{k=1,2} \Xi_k(y) + \Xi_3(y), \quad \text{say.}
\end{aligned} \tag{4.29}$$

Furthermore, for $k = 1, 2$, on applying Lemma 4.16 with some

$$J \ge \delta^{-2} \tag{4.30}$$

to be specified later (see (4.55) below), for *any* choice of summation ranges for m, n,

$$\begin{aligned}
&\left| \sum_{m,n} \sum a_m b_n \Xi_k(mn) \right| \\
&\le \sum_{\ell=0,1} \left| \sum_{m,n} \sum a_m b_n (\psi - \psi^*)(\tilde{\beta}_k + (-1)^\ell \delta - x_{k,mn}) \right| + \\
&\quad + \left| \sum_{m,n} \sum a_m b_n \sum_{\ell=0,1} (-1)^\ell \psi^*(\tilde{\beta}_k - (-1)^\ell \delta - x_{k,mn}) \right|.
\end{aligned} \tag{4.31}$$

The last sum therein is

$$\begin{aligned}
&\left| \sum_{m,n} \sum a_m b_n \sum_{1 \le |j| < J} \frac{\mathrm{e}(j\delta) - \mathrm{e}(-j\delta)}{2\pi i j} W\left(\tfrac{j}{J}\right) \mathrm{e}(j(\tilde{\beta}_k - x_{k,mn})) \right| \\
&\le \sum_{1 \le |j| < J} \frac{|\mathrm{e}(j\delta) - \mathrm{e}(-j\delta)|}{2\pi |j|} \left| \sum_{m,n} \sum a_m b_n \, \mathrm{e}(-j x_{k,mn}) \right| \\
&\le \sum_{1 \le |j| < J} \min\left\{ \frac{1}{\pi |j|}, 2\delta \right\} \left| \sum_{m,n} \sum a_m b_n \, \mathrm{e}(-j x_{k,mn}) \right|,
\end{aligned} \tag{4.32}$$

whereas, on writing B for the maximum of all $|b_n|$ with n appearing in the summation $\sum_{m,n}\sum$, the first sum on the right hand side of (4.31) may be bounded by

$$\sum_{\ell=0,1}\sum_{m,n}\sum |b_n| \cdot |(\psi - \psi^*)(\tilde{\beta}_k + (-1)^\ell \delta - x_{k,mn})|$$

$$\leq \frac{B}{J}\sum_{m,n}\sum 1 + \frac{B}{J}\sum_{1\leq|j|<J}\left|\sum_{m,n}\sum e(-jx_{k,mn})\right|.$$

Next, we have to consider

$$\left|\sum_{m,n}\sum a_m b_n \Xi_3(mn)\right|.$$

Here $\Xi_3(mn)$ is a product of two terms, each of which may be expanded into three terms as was done in (4.31):

$$\Xi_3(mn) = \prod_{k=1,2}(A_k + B_k + C_k) = \begin{aligned} &C_1C_2 + A_1C_2 + C_1A_2 + C_1B_2 + B_1C_2 + \\ &+ A_1A_2 + A_1B_2 + B_1A_2 + B_1B_2, \end{aligned}$$

where

$$A_k = \psi(\tilde{\beta}_k - \delta - x_{k,mn}) - \psi^*(\tilde{\beta}_k - \delta - x_{k,mn}),$$
$$B_k = \psi^*(\tilde{\beta}_k + \delta - x_{k,mn}) - \psi(\tilde{\beta}_k + \delta - x_{k,mn}),$$
$$C_k = \psi^*(\tilde{\beta}_k - \delta - x_{k,mn}) - \psi^*(\tilde{\beta}_k + \delta - x_{k,mn})$$

(mind the dependence on mn which we have dropped from the notation for the sake of better readability).

Along similar lines as (4.32), we obtain

$$\left|\sum_{m,n}\sum a_m b_n C_1C_2\right| \leq \sum_{\substack{1\leq|j_1|<J\\1\leq|j_2|<J}}\sum \min\left\{\frac{1}{\pi|j_1|}, 2\delta\right\}\min\left\{\frac{1}{\pi|j_2|}, 2\delta\right\} \times$$

$$\times \left|\sum_{m,n}\sum a_m b_n e(-j_1 x_{1,mn} - j_2 x_{2,mn})\right|.$$

Also, if $X_k \in \{A_k, B_k\}$ $(k = 1, 2)$, then

$$\left|\sum_{m,n}\sum a_m b_n X_k C_{3-k}\right| \tag{4.33}$$

$$\leq B \sum_{1\leq|j_2|<J}\min\left\{\frac{1}{\pi|j_2|}, 2\delta\right\}\sum_{m,n}\sum |X_k|$$

$$\leq \frac{B}{2J}\sum_{\substack{1\leq|j_1|<J\\1\leq|j_2|<J}}\sum \min\left\{\frac{1}{\pi|j_2|}, 2\delta\right\}\left|\sum_{m,n}\sum e(-j_1 x_{k,mn})\right| +$$

$$+ \frac{B}{2J}\sum_{1\leq|j_2|<J}\min\left\{\frac{1}{\pi|j_2|}, 2\delta\right\}\sum_{m,n}\sum 1.$$

If additionally $Y_{3-k} \in \{A_{3-k}, B_{3-k}\}$, then

$$\left| \sum_{m,n} \sum a_m b_n X_k Y_{3-k} \right|$$

$$\leq B \sum_{m,n} \sum |X_k Y_{3-k}|$$

$$\leq \frac{B}{4J^2} \sum_{m,n} \sum 1 + \frac{B}{4J^2} \sum_{\ell=1,2} \sum_{1 \leq |j| < J} \left| \sum_{m,n} \sum e(-jx_{\ell,mn}) \right| +$$

$$+ \frac{B}{4J^2} \sum_{\substack{1 \leq |j_1| < J \\ 1 \leq |j_2| < J}} \sum \left| \sum_{m,n} \sum e(-j_1 x_{k,mn} - j_2 x_{3-k,mn}) \right|.$$

4.5.2 Putting things together

Finally, on writing

$$\sum_{mn \in \mathcal{A}}^{*} \sum a_m b_n = \sum_{mn \in \mathcal{B}}^{*} \sum a_m b_n \mathbf{1}_{\mathcal{A}}(mn),$$

where the star in the summation indicates that the range of m is to be restricted to a Type I or Type II range, and combining in (4.29) all the estimates from the previous subsection, we may now bound

$$E = \sum_{mn \in \mathcal{A}}^{*} \sum a_m b_n - 4\delta^2 \sum_{mn \in \mathcal{B}}^{*} \sum a_m b_n. \qquad (4.34)$$

Indeed, using (4.30) and $\delta \leq \frac{1}{2}$ multiple times, and after some simplification of intermediate results,

$$\frac{|E|}{8} \leq \frac{B \log J}{J} \sum_{mn \in \mathcal{B}} \sum 1 + \qquad (4.35)$$

$$+ \delta \max_{k=1,2} \sum_{1 \leq |j| < J} \Pi_1(j) \left| \sum_{mn \in \mathcal{B}}^{*} \sum a_m b_n \, e(-jx_{k,mn}) \right| +$$

$$+ B \log J \max_{k=1,2} \sum_{1 \leq |j| < J} \Pi_2(j) \left| \sum_{mn \in \mathcal{B}}^{*} \sum e(-jx_{k,mn}) \right| +$$

$$+ \sum_{\substack{1 \leq |j_1| < J \\ 1 \leq |j_2| < J}} \sum \Pi_1(j_1, j_2) \left| \sum_{mn \in \mathcal{B}}^{*} \sum a_m b_n \, e(-j_1 x_{1,mn} - j_2 x_{2,mn}) \right| +$$

$$+ B \sum_{\substack{1 \leq |j_1| < J \\ 1 \leq |j_2| < J}} \sum \Pi_2(j_1, j_2) \left| \sum_{mn \in \mathcal{B}}^{*} \sum e(-j_1 x_{1,mn} - j_2 x_{2,mn}) \right|,$$

where

$$\Pi_r(j) = \min\{|j|^{-1}, \delta^r\}, \quad \Pi_r(j_1, j_2) = \Pi_r(j_1)\Pi_r(j_2). \tag{4.36}$$

Using Lemma A.9, we have

$$\sum_{mn \in \mathscr{B}} \sum 1 \leq \#\{\text{units in } \mathcal{O}\} \cdot \sum_{N\mathfrak{a} < x} d(\mathfrak{a}) < 6 \cdot 2^7 x (\log x)^4. \tag{4.37}$$

Remark 4.18. *The reader should compare (4.35) with [4, (10) and (12)] where the terms corresponding to the third line on the right hand side of (4.35) have a factor δ instead of our $B \log J$; this appears to be incorrect and its correction forces us to take $J \geq \delta^{-2}$ instead of just $J \geq \delta^{-1}$ as Baier does. As a consequence of having to take J larger, we end up with the decreased quality of our final result as commented on earlier in Remark 4.3.*
Furthermore, the reader should note the importance of the coefficient $\Pi_2(j)$ in the aforementioned third line. If we had written $\Pi_1(j)$ instead, then we would have ended up with an even smaller exponent in (4.5). (A rough calculation shows that one would have to replace $\frac{1}{48}$ with $\frac{1}{50}$.)

4.5.3 Removing the weights: dyadic intervals

Here we shall remove the weights (4.36) attached to some of the sums in (4.35). This may be achieved by splitting the summation over j (or j_1, j_2) into dyadic intervals: indeed, for any non-negative $f : \mathbb{Z}^2 \to \mathbb{R}$, letting

$$F(J_1, J_2) = \sum_{\substack{0 \leq |j_1| < J_1 \\ 0 \leq |j_2| < J_2 \\ (j_1, j_2) \neq (0,0)}} \sum f(j_1, j_2), \tag{4.38}$$

and writing $U = 2 \log J / \log 2$, we find that, for $r = 1, 2$,

$$\sum_{\substack{1 \leq |j_1| < J \\ 1 \leq |j_2| < J}} \sum \Pi_r(j_1, j_2) f(j_1, j_2) \leq U^2 \max_{\substack{1 \leq J_1 \leq J \\ 1 \leq J_2 \leq J}} \Pi_r(J_1, J_2) F(J_1, J_2),$$

$$\sum_{1 \leq |j| < J} \Pi_r(j) f(j, 0) \leq U \max_{1 \leq J_1 \leq J} \Pi_r(J_1) F(J_1, 1),$$

$$\sum_{1 \leq |j| < J} \Pi_r(j) f(0, j) \leq U \max_{1 \leq J_2 \leq J} \Pi_r(J_2) F(1, J_2).$$

Of course, we shall apply this with

$$f(j_1, j_2) = \left| \sum_{mn \in \mathscr{B}}^* \sum a_m b_n \, \mathrm{e}(-j_1 x_{1,mn} - j_2 x_{2,mn}) \right|, \tag{4.39}$$

and with $f(j_1, j_2)$ given by the same expression a_m and b_n replaced by 1 in order to accommodate for the fact that also such sums appear in (4.35).

Now assume for the moment that we have bounds

$$F(J_1, J_2) \leq \mathcal{F}(J_1, J_2),$$

where the left hand side is symmetric in both arguments and does not depend on the particular choice of the coefficients in (4.39) (but, of course, still subject to the Type I/II conditions presented in Section 4.5.1); The reader may wish to glance at Proposition 4.20 and Proposition 4.21 below, where we furnish such bounds for the Type II and Type I sums respectively. Then, using (4.35), Lemma A.8 and (4.37), we have

$$|E| < 2^{14}\epsilon^{-2}M_{\mathcal{O}}(\epsilon)(xJ)^{\epsilon}\left(\epsilon^{-3}x^{1+\epsilon}J^{-1} + \right. \tag{4.40}$$
$$+ \max_{1 \leq J_1 \leq J} \max\{\delta\Pi_1(J_1), \Pi_2(J_1)\}\mathcal{F}(J_1, 1) +$$
$$\left. + \max_{\substack{1 \leq J_1 \leq J \\ 1 \leq J_2 \leq J}} \Pi_1(J_1, J_2)\mathcal{F}(J_1, J_2)\right).$$

4.5.4 Transforming the argument in the exponential term

In the proof of the bounds for the Type I and Type II sums we need to combine variables in $\mathcal{O}$ (see Sections 4.5.5 and 4.5.6 below). Having this goal in mind, the shape of the argument of the exponential in (4.39) appears to be, at a superficial glance, a technical obstruction.

However, this putative problem vanishes after a simple variable transformation that we shall now describe: by definition of $x_{k,mn}$,

$$-j_1 x_{1,mn} - j_2 x_{2,mn} = -j_1 \mathfrak{R}_{\omega}(mn\vartheta) - j_2 \mathfrak{I}_{\omega}(mn\vartheta). \tag{4.41}$$

Letting ξ_2 be given as in (4.11) and writing $\ell = \ell_1 + \ell_2\omega$, a short computation yields

$$\mathfrak{I}_{\omega}(\ell\rho) = \ell_2 \mathfrak{R}_{\omega}\rho + (\ell_1 + \ell_2\xi_2)\mathfrak{I}_{\omega}\rho \quad (\rho \in \mathbb{C}).$$

Then, via the equivalence

$$\begin{pmatrix} 0 & -1 \\ -1 & -\xi_2 \end{pmatrix}\begin{pmatrix} \ell_1 \\ \ell_2 \end{pmatrix} = \begin{pmatrix} j_1 \\ j_2 \end{pmatrix} \iff \begin{pmatrix} \ell_1 \\ \ell_2 \end{pmatrix} = \begin{pmatrix} \xi_2 & -1 \\ -1 & 0 \end{pmatrix}\begin{pmatrix} j_1 \\ j_2 \end{pmatrix}, \tag{4.42}$$

and assuming $(j_1, j_2) \neq (0,0)$, we observe that (4.41) equals $\mathfrak{I}_{\omega}(\ell mn\vartheta)$ when (ℓ_1, ℓ_2) is calculated via the above formula. We let $\mathscr{L}(J_1, J_2)$ be the set of algebraic integers $\ell \in \mathcal{O}$ arising from (j_1, j_2) via the above formula, that is, $\mathscr{L}(J_1, J_2)$ is the set

$$\{(\xi_2 j_1 - j_2) - j_1\omega : |j_1| < J_1, |j_2| < J_2, (j_1, j_2) \neq (0,0)\}.$$

Consequently, if $F(J_1, J_2)$ is given by (4.38) with f given by (4.39), then

$$F(J_1, J_2) = \sum_{\ell \in \mathscr{L}(J_1, J_2)}\left|\sum_{mn \in \mathscr{B}}^{*}\sum a_m b_n \, e(\mathfrak{I}_{\omega}(\ell mn\vartheta))\right|. \tag{4.43}$$

50

For a later extension of the summation over ℓ, we note that $\mathscr{L}(J_1, J_2)$ is contained in the set of all ℓ satisfying

$$1 \leq N(\ell) < 5|\omega|^4 (J_1 + J_2)^2. \tag{4.44}$$

The reader will note that this set potentially contains many more elements than $\mathscr{L}(J_1, J_2)$, for we obviously have

$$\#\mathscr{L}(J_1, J_2) \leq (2J_1 + 1)(2J_2 + 1) \leq 9\,J_1 J_2; \tag{4.45}$$

in any case, we require both Eqs. (4.44) and (4.45).

Remark 4.19. *The reader may ponder about the reason for our bias towards using $\Im_\omega$ instead of $\Re_\omega$. In fact, this is no coincidence, for we have*

$$\Re_\omega(\ell\rho) = \ell_1 \Re_\omega \rho + \ell_2 (\Re_\omega \omega^2) \Im_\omega \rho \quad (\rho \in \mathbb{C})$$

and an analogue of (4.42) is not possible due to failure of the matrix coefficients to be integral if $|\Re_\omega \omega^2| > 1$ or even failure of the matrix to be invertible if $\Re_\omega \omega^2 = 0$ (consider, for instance, the natural choice $\omega = i$).

4.5.5 The Type II sums

Proposition 4.20 (Type II bound). *Consider F from (4.38) with f given by (4.39) subject to*

$$\sum_{mn \in \mathscr{B}}^{*} \sum = \sum_{\substack{x/2 \leq N(mn) < x \\ x^\mu < N(m) < x^{\mu+\kappa}}} \sum,$$

where $\mu \in (0,1]$, $\kappa \in (0, \frac{1}{2}]$ and $x \geq 3$. For the coefficients in (4.39) assume that $|a_m| \leq 1$ and $|b_n| \leq d((n))$. Moreover, suppose that a, q, γ and C are as in (4.4) and let $M_\mathcal{O}(\epsilon)$ be as in Lemma A.8. Then, for any $\epsilon \in (0, \frac{1}{2}]$,

$$F(J_1, J_2) < 2^{21} C |\omega|^7 \epsilon^{-5} M_\mathcal{O}(\epsilon)^{3/2} x^{3\epsilon} \times \left(J_1 J_2 x^{(1+\mu+\kappa)/2} + (J_1 J_2)^{1/2+\epsilon} \times \right. \tag{4.46}$$
$$\left. \times \left((J_1 + J_2) x |q|^{-1} + (J_1 + J_2) x^{1-\mu/4} + |q| x^{(2+\mu+\kappa)/4} \right) \right).$$

Proof. Looking at (4.38), we may split the summation over m into "dyadic annuli," getting

$$F(J_1, J_2) \leq (\mu + \kappa) \frac{\log x}{\log 2} \max_{\substack{x^\mu < K, K' \leq x^{\mu+\kappa} \\ K \leq K' < 2K}} F(J_1, J_2, K, K'), \tag{4.47}$$

where, upon employing the transformation described in Section 4.5.4 along the way, $F(J_1, J_2, K, K')$ may be taken to be

$$\sum_{\substack{\ell \in \mathscr{L}(J_1, J_2) \\ K \leq N(m) < K'}} \sum \left| \sum_{x/2 \leq N(nm) < x} b_n \, \mathrm{e}(\Im_\omega(\ell mn\vartheta)) \right|.$$

51

(Here and in the following we are always assuming J_1, J_2, K and K' to be positive integers such that $K \le K' < 2K$.) By (4.45) and Lemma A.3,

$$\sum_{\substack{\ell \in \mathscr{L}(J_1, J_2) \\ K \le N(m) < K'}} \sum 1 \le 13 \cdot 9\, J_1 J_2 K'.$$

Hence, letting

$$Q = \frac{F(J_1, J_2, K, K')^2}{13 \cdot 9\, J_1 J_2 K'}, \tag{4.48}$$

Cauchy's inequality gives

$$Q \le \sum_{\substack{\ell \in \mathscr{L}(J_1, J_2) \\ K \le N(m) < K'}} \left| \sum_{x/2 \le N(nm) < x} b_n\, \mathrm{e}(\Im_\omega(\ell m n \vartheta)) \right|^2,$$

which, upon expanding the square and rearranging, yields

$$Q \le \sum_{\substack{\ell \in \mathscr{L}(J_1, J_2) \\ x/2K \le N(n) < x/K' \\ x/2K \le N(\tilde{n}) < x/K'}} \sum \sum |b_n \overline{b_{\tilde{n}}}| \left| \sum_m^* \mathrm{e}(\Im_\omega(\ell m (n - \tilde{n})\vartheta)) \right|,$$

where $\sum_m^*$ restricts the summation to those m with

$$\max\{K, x/2N(n), x/2N(\tilde{n})\} \le N(m) < \min\{K', x/N(n), x/N(\tilde{n})\}.$$

Next, we isolate the "diagonal contribution" Δ, that is, those terms where $n = \tilde{n}$, for in this case the sum over m can only be bounded trivially. Using Lemma A.9, (4.45) and Lemma A.3, this is found to be

$$
\begin{aligned}
\Delta &= \sum_{x/2K \le N(n) < x/K'} |b_n|^2 \sum_{\ell \in \mathscr{L}(J_1, J_2)} \sum_m^* 1 \\
&< (6 \cdot 2^{10} x (K')^{-1} (\log x)^8)(9\, J_1 J_2)(13\, K') \\
&< 2^{20} J_1 J_2 x (\log x)^8.
\end{aligned} \tag{4.49}
$$

Moreover, using Lemma A.8 and Lemma 4.9, we have

$$Q \le \Delta + 13 |\omega|^2 M_{\mathcal{O}}(\epsilon)^2 (x/K')^{2\epsilon} \sqrt{K'} \sum_{1 \le N(j) < K_{\mathrm{II}}} c_j E(j, \sqrt{K'}),$$

where E is given by (4.17),

$$K_{\mathrm{II}} = 20 |\omega|^4 (J_1 + J_2)^2 x K^{-1}$$

and

$$c_j = \sum_{\substack{\ell \in \mathscr{L}(J_1, J_2) \\ \ell \mid j}} \sum_{\substack{x/2K \le N(n) < x/K' \\ x/2K \le N(\tilde{n}) < x/K' \\ j/\ell = (n - \tilde{n})}} 1 \le M_{\mathcal{O}}(\epsilon) K_{\mathrm{II}}^{\epsilon} \cdot (13\, x/K').$$

Thus, using (4.49), Theorem 4.11 and recalling (4.48),

$$F(J_1, J_2, K, K')^2 < 2^{28}(J_1 J_2)^2 x K (\log x)^8 + 2^{20} \cdot 13^3 \cdot 9 \,|\omega|^2 M_{\mathcal{O}}(\epsilon)^3 \times$$
$$\times (x/K')^{2\epsilon} K_{\mathrm{II}}^{\epsilon} J_1 J_2 x \sqrt{K'} \times$$
$$\times (1 + C^2 |\omega|^4 K_{\mathrm{II}} |q|^{-2})(\sqrt{K'} + |\omega|^2 |q|^2 \log(2\sqrt{K'})).$$

Upon taking the square root, and simplifying the resulting expressions,

$$F(J_1, J_2, K, K') < 2^{21} C |\omega|^7 \epsilon^{-4} M_{\mathcal{O}}(\epsilon)^{3/2} x^{2\epsilon} \big(J_1 J_2 \sqrt{xK} + (J_1 J_2)^{1/2 + \epsilon} \times$$
$$\times \big((J_1 + J_2) x |q|^{-1} + (J_1 + J_2) x K^{-1/4} + |q| x^{1/2} K^{1/4}\big)\big).$$

Recalling (4.47), we infer (4.46). □

4.5.6 The Type I sums

Proposition 4.21 (Type I bound). *Consider F from (4.38) with f given by (4.39) subject to*

$$\sideset{}{^*}\sum_{mn \in \mathscr{B}} \sum = \sum_{\substack{x/2 \le N(mn) < x \\ N(m) < M}} \sum,$$

where $M \le x$ and $x \ge 3$. For the coefficients in (4.39) assume that $|a_m| \le 1$ and $b_n = \mathbf{1}_{\{1 \le N(n) < x\}}$. Moreover, suppose that a, q, γ and C are as in (4.4) and let $M_{\mathcal{O}}(\epsilon)$ be as in Lemma A.8. Then, for any $\epsilon \in (0, \frac{1}{2}]$,

$$F(J_1, J_2) \le 2^{38} C^2 |\omega|^{14} \epsilon^{-2} M_{\mathcal{O}}(\epsilon)(J_1 + J_2)^{2\epsilon} (x^2 |q|)^{\epsilon} \times \qquad (4.50)$$
$$\times \big((J_1 + J_2)^2 x |q|^{-2} + (J_1 + J_2)^2 x^{1/2} M^{1/2} + x^{1/2} |q|^2\big).$$

Proof. As we did with the Type II sums in the proof of Proposition 4.20, we may split the summation over m into dyadic annuli, getting

$$F(J_1, J_2) \le \frac{\log x}{\log 2} \max_{\substack{1 \le K, K' \le M \\ K \le K' < 2K}} \tilde{F}(J_1, J_2, K, K'), \qquad (4.51)$$

where $\tilde{F} = \tilde{F}(J_1, J_2, K, K')$ is given by

$$\sum_{\substack{\ell \in \mathscr{L}(J_1, J_2) \\ K \le N(m) < K'}} \sum \Big| \sum_{x/2 \le N(nm) < x} \mathrm{e}(\Im_{\omega}(\ell m n \vartheta)) \Big|.$$

53

By Lemma 4.9 and Lemma A.8,

$$\tilde{F} \le 13|\omega|^2 \sqrt{x/K} \sum_{\substack{\ell \in \mathscr{L}(J_1, J_2) \\ K \le N(m) < K'}} \sum E(\ell m, \sqrt{x/K})$$

$$\le 13|\omega|^2 M_{\mathcal{O}}(\epsilon) K_{\mathrm{I}}^{\epsilon} \sqrt{x/K} \sum_{K \le N(j) < K_{\mathrm{I}}} E(j, \sqrt{x/K}),$$

with E given by (4.17) and

$$K_{\mathrm{I}} = 10|\omega|^4 (J_1 + J_2)^2 K. \tag{4.52}$$

Theorem 4.11 now shows that

$$\tilde{F} \le 13 \cdot 2^{20} M_{\mathcal{O}}(\epsilon) K_{\mathrm{I}}^{\epsilon} |\omega|^2 \sqrt{x/K} (1 + C^2 |\omega|^4 K_{\mathrm{I}} |q|^{-2}) \times$$
$$\times (\sqrt{x/K} + |\omega|^2 |q|^2 \log(2\sqrt{x/K}))$$
$$\le 2^{26} |\omega|^4 M_{\mathcal{O}}(\epsilon)(J_1 + J_2)^{2\epsilon} x^{\epsilon} (xK^{-1}) +$$
$$+ 2^{28} C^2 |\omega|^{14} \epsilon^{-1} M_{\mathcal{O}}(\epsilon)(J_1 + J_2)^{2\epsilon} x^{\epsilon} \times$$
$$\times ((J_1 + J_2)^2 x|q|^{-2} + (J_1 + J_2)^2 x^{1/2} K^{1/2} + x^{1/2} |q|^2 K^{-1/2}).$$

Herein, for very small K, the term xK^{-1} becomes problematic. To circumvent this, we note that Theorem 4.11 also furnishes the bound

$$\tilde{F} \le 2^{20} |\omega|^7 \epsilon^{-1} M_{\mathcal{O}}(\epsilon)(J_1 + J_2)^{2\epsilon}(x|q|)^{\epsilon} x^{1/2} |q|^2 K^{-1/2},$$

provided that $K_{\mathrm{I}} \le |q|^2/(12C|\omega|^2)^2$. On the other hand, if $K_{\mathrm{I}} > |q|^2/(12C|\omega|^2)^2$, then, recalling (4.52), we have

$$xK^{-1} = 10|\omega|^4 (J_1 + J_2)^2 xK_{\mathrm{I}}^{-1} < 2^{11} |\omega|^8 C^2 (J_1 + J_2)^2 x|q|^{-2}.$$

Therefore, after joining both bounds,

$$\tilde{F} \le 2^{38} C^2 |\omega|^{14} \epsilon^{-1} M_{\mathcal{O}}(\epsilon)(J_1 + J_2)^{2\epsilon}(x|q|)^{\epsilon} \times$$
$$\times ((J_1 + J_2)^2 x|q|^{-2} + (J_1 + J_2)^2 x^{1/2} K^{1/2} + x^{1/2} |q|^2 K^{-1/2}).$$

Upon plugging this into (4.51), we obtain (4.50). $\qquad\square$

4.5.7 Assembling the parts

Finally, we are in a position to use (4.40). Assume the hypotheses of Proposition 4.20. Then, looking at (4.46), we use (4.40) together with

$$\Pi_r(H) = \min\{H^{-1}, \delta^r\} \le \min\{H^{-1}, \delta^{r/2} H^{-1/2}\} \tag{4.53}$$

and the corresponding bound for $\Pi(J_1, J_2)$, where we use the first term on the right hand side of (4.53) for the diagonal contribution $J_1 J_2 x^{(1+\mu+\kappa)/2}$ and the second term for the rest. The result for the error in the Type II sums (see (4.34)) is

$$\frac{\epsilon^5 |E|}{2^{39} C |\omega|^7 M_{\mathcal{O}}(\epsilon)^{5/2}} < (x^2 J)^{\epsilon} x J^{-1} + (x^4 J^3)^{\epsilon} \times$$
$$\times (x^{(1+\mu+\kappa)/2} + \delta J x|q|^{-1} + \delta J x^{1-\mu/4} + \delta |q| x^{(2+\mu+\kappa)/4}).$$

Moving on to the Type I, assuming the hypotheses of Proposition 4.21 and using $\Pi_r(H) \leq \delta^r$ in (4.40), we infer the following error estimate for the Type I sums:

$$\frac{|E|}{2^{58}C^2|\omega|^{14}\epsilon^{-5}M_{\mathcal{O}}(\epsilon)^2} < (x^2 J)^\epsilon x J^{-1} + (x^3 J^3 |q|)^\epsilon \times$$
$$\times (\delta^2 J^2 x |q|^{-2} + \delta^2 J^2 x^{1/2} M^{1/2} + \delta^2 x^{1/2}|q|^2).$$

On recalling (4.34) and plugging the above bounds into Theorem 4.7, we find that the error

$$\tilde{E} = \frac{S(\mathscr{A}, x^\kappa) - 4\delta^2 S(\mathscr{B}, x^\kappa)}{2^{60}C^2|\omega|^{14}\epsilon^{-8}M_{\mathcal{O}}(\epsilon)^{5/2}}$$

satisfies the bound

$$|\tilde{E}| < (x^3 J)^\epsilon x J^{-1} +$$
$$+ (x^3 J^3 |q|)^\epsilon (\delta^2 J^2 x |q|^{-2} + \delta^2 J^2 x^{1/2} M^{1/2} + \delta^2 x^{1/2}|q|^2) +$$
$$+ (x^5 J^3)^\epsilon (x^{(1+\mu+\kappa)/2} + \delta J x |q|^{-1} + \delta J x^{1-\mu/4} + \delta|q|x^{1/2+(\mu+\kappa)/4}).$$

Evidently, this bound is increasing with κ and to detect primes, we must take $\kappa = \frac{1}{2}$. Since

$$S(\mathscr{B}, x^{1/2}) \asymp \frac{x}{\log x}$$

(see Corollary A.6 and the discussion that follows it), we shall aim for a bound of the type

$$|\tilde{E}| \ll_{C,\omega,\varpi} (\delta^2 x)x^{-\varpi} \tag{4.54}$$

with some exponent $\varpi > 0$ and δ in some range (w.r.t. x) as large as possible. Furthermore, we must satisfy (4.30). With these constraints in mind, and given q, we take $x = |q|^{12}$ (as was stated in Theorem 4.4 anyway) so that $|q| = x^{1/12}$ and, moreover,

$$J = \lceil \delta^{-2} x^{5\epsilon} \rceil, \quad M = x^{5/6}, \quad \mu = \frac{1}{3}. \tag{4.55}$$

Then, on writing $\delta = x^{-d}$ $(d > 0)$, we find

$$|\tilde{E}| < \delta^2 x \cdot 14(x^{\epsilon(-2+2d+5\epsilon)} + x^{-1/12+4d+\epsilon(157/12+6d+15\epsilon)}).$$

For the last bound to be $o(1)$, the last term in the parentheses certainly forces $d < \frac{1}{48}$. Using this, the first term in the parentheses is $\leq x^{-\epsilon}$ provided that $\epsilon \leq \frac{23}{120} = 0.19\ldots$; moreover, the second term in the parentheses is $\leq x^{-\epsilon}$ if

$$\epsilon \leq \frac{\sqrt{29281}-169}{360} \quad \text{and} \quad d \leq \frac{1-169\epsilon-180\epsilon^2}{48+72\epsilon}$$

and the latter condition is definitely satisfied if $d \leq \frac{1}{48} - 4\epsilon$.

Therefore,

$$|\tilde{E}| < 28\delta^2 x^{1-\epsilon} \quad \text{if } x^{-1/48+4\epsilon} \leq \delta \leq \frac{1}{2}.$$

This concludes the proof of Theorem 4.4.

55

Chapter 5

Beatty-type sets in imaginary quadratic number fields

Let $\mathbb{K}$, $\mathcal{O}$ and ω from Chapter 4 retain their meaning. In this chapter, we shall be interested in the Beatty-type sets $\mathcal{B}_\omega(\alpha, \beta)$ as defined in (1.10). In Section 5.1 we direct attention to an interesting phenomenon with respect to the natural density of $\mathcal{B}_\omega(\alpha, \beta)$ which has no matching analogue in the realm of ordinary Beatty sets as defined in (1.2). After that, we investigate the existence of prime elements in $\mathcal{B}_\omega(\alpha, \beta)$, applying the results from Chapter 4 in the process. Using the Hurwitz continued fraction algorithm for $\mathbb{Z}[i]$, we obtain an analogue of Theorem 1.6 for the Gaussian integers.

5.1 Density

We preface the subsequent investigation with a short comment concerning the one-dimensional case. To this end, consider an ordinary two-sided Beatty set $\mathcal{B}_{ts}(\alpha, \beta)$ as defined in (1.3). (One-sided ordinary Beatty sets $\mathcal{B}(\alpha, \beta)$ can be treated similary, but, as already mentioned in the introduction, we favour contrasting our number field analogue with two-sided Beatty sets.) For $|\alpha| \leq 1$ we have $\mathcal{B}_{ts}(\alpha, \beta) = \mathbb{N}$, so we shall assume that $|\alpha| > 1$. Then $\mathcal{B}_{ts}(\alpha, \beta)$ contains

$$2\alpha^{-1}x(1 + O(1/x)) \tag{5.1}$$

many elements m with $|m| \leq x$, as can be seen quite easily by noting that, for integers $n \neq \tilde{n}$, the numbers $n\alpha + \beta$ and $\tilde{n}\alpha + \beta$ have distance $\geq \alpha \geq 1$ and, therefore, yield different integers after applying the floor function to them; the number of integers n satisfying $|\lfloor n\alpha + \beta \rfloor| \leq x$ is immediately seen to be (5.1), and, by the previous argument, equals $\#\{m \in \mathcal{B}_{ts}(\alpha, \beta) : |m| \leq x\}$. In particular, the *natural density* (or simply *density*)

$$\text{dens}\, \mathcal{B}_{ts}(\alpha, \beta) = \lim_{x \to \infty} \frac{\#\{m \in \mathcal{B}_{ts}(\alpha, \beta) : |m| \leq x\}}{\#\{m \in \mathbb{Z} : |m| \leq x\}}$$

exists and equals α^{-1}.

Turning to Beatty-type sets $\mathcal{B}_\omega(\alpha, \beta)$ with $\alpha, \beta \in \mathbb{C}$, the corresponding analysis of the limit

$$\text{dens}\, \mathcal{B}_\omega(\alpha, \beta) = \lim_{x \to \infty} \frac{\#\{m \in \mathcal{B}_\omega(\alpha, \beta) : N(m) \leq x\}}{\#\{m \in \mathcal{O} : N(m) \leq x\}} \tag{5.2}$$

turns out to be more subtle. As an illustration of the statement, consider the following:

Example 5.1. *Let* $\alpha = \frac{8}{7}\exp(\frac{2\pi i}{3})$. *Then* $|\alpha| = \frac{8}{7} > 1$, *but* $\lfloor 0\alpha \rfloor_i = 0 = \lfloor -i\alpha \rfloor_i$; *the representation of* $0 \in \mathcal{B}_i(\alpha, 0)$ *as* $\lfloor n\alpha \rfloor_i$ *(*$n \in \mathbb{Z}[i]$*) is not unique. On the other hand, obviously elements in* $\mathcal{B}_i(1,0)$ *are represented uniquely.*

As of the writing of this thesis, our insights concerning existence and value of the limit (5.2) are quite superficial and it seems desirable to gain a more complete understanding here. Nonetheless, we are able to compute the limit in (5.2) as $|\alpha|^{-2}$ provided that $|\alpha|$ is not "too small" in terms of the geometry of the fundamental parallelogram Λ given by (4.6) and the lattice $\mathcal{O}$: denote the *diameter* of Λ by

$$\mathrm{diam}\,\Lambda = \sup\{|x - y| : x, y \in \Lambda\},$$

and let

$$\lambda(\mathcal{O}) = \min\{|x - y| : x, y \in \mathcal{O}, x \neq y\}.$$

Our result may now be stated as follows:

Proposition 5.2. *Assuming*

$$|\alpha|\lambda(\mathcal{O}) \geq \mathrm{diam}\,\Lambda, \tag{5.3}$$

we have

$$\mathrm{dens}\,\mathcal{B}_\omega(\alpha, \beta) = |\alpha|^{-2}. \tag{5.4}$$

The proof of Proposition 5.2 is a generalisation of the argument presented above for ordinary Beatty sets. We need the following lemma:

Lemma 5.3. *Assuming (5.3), the representation of* $m \in \mathcal{B}_\omega(\alpha, \beta)$ *as* $m = \lfloor n\alpha + \beta \rfloor_\omega$ *with* $n \in \mathcal{O}$ *is unique.*

Proof. Note that the points $n\alpha + \beta$ ($n \in \mathcal{O}$) are spaced by at least $|\alpha|\lambda(\mathcal{O})$ apart. Upon observing that $\lfloor x \rfloor_\omega = \lfloor \tilde{x} \rfloor_\omega$ if and only if there is some $k \in \mathcal{O}$ such that x and $\tilde{x}$ both lie in the shifted fundamental parallelogram $k + \Lambda$, we conclude that (5.3) ensures that no two distinct points $n\alpha + \beta$ ($n \in \mathcal{O}$) map to the same value under $\lfloor \cdot \rfloor_\omega$. $\qquad\square$

Example 5.4. *If* $\omega = i$, *then (5.3) takes the form* $|\alpha| \geq \sqrt{2}$. *The* α *exhibited in Example 5.1 fails to satisfy this condition. However, the last part in Example 5.1 shows that (5.3) is not necessary for the conclusion of Lemma 5.3 to hold.*

Proof of Proposition 5.2. By Lemma 5.3, (5.3) guarantees that

$$\#\{m \in \mathcal{B}_\omega(\alpha, \beta) : N(m) \leq x\} = \#\{n \in \mathcal{O} : |\lfloor n\alpha + \beta \rfloor_\omega| \leq \sqrt{x}\}.$$

Denoting the latter set by $\mathscr{X}_\omega(\alpha, \beta; \sqrt{x})$, and using

$$\big|\,|\lfloor n\alpha + \beta \rfloor_\omega| - |n\alpha + \beta|\,\big| < \mathrm{diam}\,\Lambda,$$

we have

$$\mathscr{X}(\alpha, \beta; \sqrt{x} - \mathrm{diam}\,\Lambda) \subseteq \mathscr{X}_\omega(\alpha, \beta; \sqrt{x}) \subseteq \mathscr{X}(\alpha, \beta; \sqrt{x} + \mathrm{diam}\,\Lambda),$$

58

where

$$\mathscr{X}(\alpha, \beta; y) = \{n \in \mathcal{O} : |n\alpha + \beta| \leq y\}.$$

From

$$\mathscr{X}(\alpha, \beta; \sqrt{x} - \Delta) \supseteq \{n \in \mathcal{O} : |n| \leq (\sqrt{x} - \Delta - |\beta|)/|\alpha|\},$$
$$\mathscr{X}(\alpha, \beta; \sqrt{x} + \Delta) \subseteq \{n \in \mathcal{O} : |n| \leq (\sqrt{x} + \Delta + |\beta|)/|\alpha|\}$$

with $\Delta = \operatorname{diam} \Lambda$ and Lemma A.2, we infer

$$\#\mathscr{X}_\omega(\alpha, \beta; \sqrt{x}) \leq \frac{\pi(\sqrt{x} + \Delta + |\beta|)^2}{|\alpha|^2 |\Im\omega|}(1 + o(1))$$

and a corresponding lower bound. Now (5.4) follows from

$$\lim_{x \to \infty} \left(\frac{\pi(\sqrt{x} \pm (\Delta + |\beta|))^2}{|\alpha|^2 |\Im\omega|} \Big/ \frac{\pi x}{|\Im\omega|} \right) = |\alpha|^{-2}. \qquad \square$$

5.2 Prime elements in Beatty-type sets

We turn to the following question: *does $\mathcal{B}_\omega(\alpha, \beta)$ contain infinitely many prime elements?* If α is not contained in $\mathbb{K}$ and $\mathbb{K}$ is assumed to have class number 1, then the results of Chapter 4 can be used to give an affirmative answer:

Theorem 5.5. *Let $\mathbb{K}$ have class number 1. Moreover, suppose that α and β are complex numbers with α not contained in $\mathbb{K}$. Then $\mathcal{B}_\omega(\alpha, \beta)$ contains infinitely many prime elements.*

The results in Chapter 4 rest on the Diophantine approximability of the number ϑ to which they are applied (see Theorem 4.4 and, in particular, the assumption (4.4) made therein). Consequently, one may state "more explicit" versions of these results if one can exhibit admissible a, q as in (4.4) more explicitly. We shall do this in terms of continued fractions. Of course, it stands to reason whether such expansions indeed match our intent of generating admissible a, q in an "explicit" fashion, but the fact that they can be computed algorithmically from an input matching ϑ with sufficient floating point precision, should at least convince the reader that working with continued fractions at this point is not completely misguided.

In any case, the particular continued fraction algorithm we shall employ here is due to Hurwitz [43] and may be described as follows: for $\vartheta^{(0)} = \vartheta \neq 0$ determine the unique $x^{(0)} \in \mathcal{O}$ such that

$$\vartheta^{(0)} - x^{(0)} \in \left\{\rho_1 + \rho_2 i \in \mathbb{C} : \rho_1, \rho_2 \in [-\tfrac{1}{2}, \tfrac{1}{2})\right\}, \qquad (5.5)$$

put

$$\vartheta^{(1)} = (\vartheta^{(0)} - x^{(0)})^{-1}$$

and proceed to compute $x^{(1)}$, $\vartheta^{(2)}$, $x^{(2)}$, and so on, in the same fashion. This yields an expansion

$$\vartheta: x^{(0)} + \cfrac{1}{x^{(1)} + \cfrac{1}{x^{(2)} + \cfrac{1}{\ddots}}}. \tag{5.6}$$

If $\vartheta \in \mathbb{Q}(i)$, then eventually

$$\vartheta^{(r)} - x^{(r)} = 0 \tag{5.7}$$

for some $r \in \mathbb{N}$, thus barring one from computing $x^{(r+1)}$; one obtains a *finite* continued fraction expansion for ϑ. In this case, we write R for the r in (5.7).

If $\vartheta \in \mathbb{C} \setminus \mathbb{Q}(i)$, then $\vartheta^{(r)} - x^{(r)} \neq 0$ for all $r \in \mathbb{N}_0$, so that $x^{(r+1)}$ is always guaranteed to exist and, as Hurwitz showed, the colon in (5.6) can be replaced with an equality if the right hand side is understood as

$$\lim_{r \to \infty} [x^{(0)}; x^{(1)}, \ldots, x^{(r)}],$$

where

$$[x^{(0)}; x^{(1)}, \ldots, x^{(r)}] = x^{(0)} + \cfrac{1}{x^{(1)} + \cfrac{1}{\ddots + \cfrac{1}{x^{(r)}}}}.$$

In this case, we put $R = \infty$.

Now compute Gaussian integers $a^{(r)}$ and $q^{(r)}$ ($r = 0, 1, 2, \ldots$ so long as $r \leq R$) according to the recurrence

$$\begin{cases} a^{(-1)} = 1, & a^{(0)} = x^{(0)}, & a^{(r)} = x^{(r)} a^{(r-1)} + a^{(r-2)}, \\ q^{(-1)} = 0, & q^{(0)} = 1, & q^{(r)} = x^{(r)} q^{(r-1)} + q^{(r-2)}. \end{cases}$$

Proposition 5.6. *With the above notation, the following four assertions hold for $r = 0, 1, 2, \ldots$ ($r \leq R$):*

- $a^{(r)}$ *and* $q^{(r)}$ *are coprime Gaussian integers.*

- $|q^{(r-1)}| < |q^{(r)}|$.

- $a^{(r)}/q^{(r)} = [x^{(0)}; x^{(1)}, \ldots, x^{(r)}]$.

- $|\vartheta - a^{(r)}/q^{(r)}| \leq (2 + \sqrt{2})|q^{(r)}|^{-2}$.

Proof. All of this is proved in [43], the last statement being implicit: combine [43, (17)] with the Hurwitz's last remark before Section I. $\square$

With this notation we may state the following result which may be viewed as an analogue of Theorem 1.6:

Theorem 5.7. *Let $\omega \in \mathbb{Z}[i]$ be such that $\mathbb{Z}[i] = \mathbb{Z}[\omega]$ and let α and β be complex numbers. Suppose that α is not contained in $\mathbb{Q}(i)$ and exceeds 1 in absolute value. Assume that $\vartheta = \alpha^{-1}$ has the Hurwitz continued fraction expansion (5.6) with $q^{(r)}$ be given as above. Then there is a prime element $p \in \mathbb{Z}[i] \cap \mathcal{B}_\omega(\alpha, \beta)$ such that*

$$N(p) \leq |q^{(r)}|^{12}, \tag{5.8}$$

where

$$r = \lceil \max\{2^{27} \log(2|\omega|) \log M_{\mathbb{Z}[i]}(\tfrac{1}{384}), \, 2^6 \log(\delta^{-1})\} \rceil$$

with δ given by (5.14) below and $M_{\mathbb{Z}[i]} = M_{\mathcal{O}}$ from Lemma A.8.

Some remarks are in order:

Remark 5.8. *The above result could also easily have been formulated in terms of the Hurwitz continued fraction expansion of α instead of α^{-1}, thus replacing $q^{(r)}$ in (5.8) with the numerator $a^{(r)}$ of a convergent to α, in turn increasing the similarity to Theorem 1.4 and Theorem 1.6.*

Remark 5.9. *For some positive integer $k \geq 18$ let $\rho(k) = \pi/k$ and consider $\mathcal{B}_i(\alpha, \beta)$ with $\alpha = 10 + \rho(k) \in \mathbb{C} \setminus \mathbb{Q}(i)$ and $\beta = 4 + 4i$. Clearly, by Theorem 5.5, there are infinitely many primes in $\mathcal{B}_i(\alpha, \beta)$ and Theorem 5.7 can be applied to bound*

$$\min\{N(p) : prime \ p \in \mathcal{B}_i(\alpha, \beta)\}. \tag{5.9}$$

For any $n \in \mathbb{Z}[i]$, we have

$$\lfloor n\alpha + \beta \rfloor_i = 10n + (4 + 4i) + \lfloor n\rho(k) \rfloor_i$$

and if $N(n) < \rho(k)^{-2}$, then

$$\lfloor n\rho(k) \rfloor_i \in \{0, -1, -1 - i, -i\}.$$

Hence, for those n, we have

$$\lfloor n\alpha + \beta \rfloor_i \equiv g(n) \mod 10\mathbb{Z}[i]$$

with some $g(n) \in \mathscr{G} = \{4 + 4i, 3 + 4i, 3 + 3i, 4 + 3i\}$. It is easily verified that none of the residue classes $\tilde{g} \mod 10\mathbb{Z}[i]$ with $\tilde{g} \in \mathscr{G}$ contains any Gaussian prime. Therefore, any $m \in \mathcal{B}_i(\alpha, \beta)$ with $N(n) < (k/\pi - |\beta|)^2$ is composite.
Since, regardless of the choice of k, we have $|\alpha| < 11$, the above conclusion shows that a general bound for (5.9) cannot depend on β and the size of α alone, in the sense that there is no function taking only $\lceil \alpha \rceil$ and β as argument and always yielding a bound for (5.9).

5.3   Detecting membership

Before proving our main results we need an analogue of Lemma 1.2. This is furnished by the following result:

Lemma 5.10. *Let* $\mathcal{B}_\omega(\alpha, \beta)$ *be a Beatty-type set as in (1.10). Then, for any algebraic integer* $m \in \mathcal{O}$,

$$m \in \mathcal{B}_\omega(\alpha, \beta) \quad \text{if and only if} \quad \frac{m}{\alpha} \in \mathscr{F} \bmod \mathcal{O}, \tag{5.10}$$

where $\mathscr{F} \bmod \mathcal{O}$ *denotes the union of all* $\mathcal{O}$*-translates of the set* $\mathscr{F} = \alpha^{-1}(\beta - \Lambda) \subseteq \mathbb{C}$ *and* Λ *is the fundamental parallelogram as given in (4.6).*

Proof. Notice that $m \in \mathcal{B}_\omega(\alpha, \beta)$ is equivalent to the existence of some $n \in \mathcal{O}$ such that $n\alpha + \beta \in m + \Lambda$. Since the latter is equivalent to $\alpha^{-1}m - n \in \alpha^{-1}(\beta - \Lambda)$, the claim follows after reducing modulo $\mathcal{O}$. $\square$

If α is real with $\alpha < -1$, then it is easily seen that there is some shift $\tilde{\beta} \in \mathbb{C}$ such that the set $\mathscr{F} \bmod \mathcal{O}$ from Lemma 5.10 satisfies

$$\mathscr{F}(-\tilde{\epsilon}) \subset (\mathscr{F} \bmod \mathcal{O}) \subset \mathscr{F}(\tilde{\epsilon}) \tag{5.11}$$

for every $\tilde{\epsilon} > 0$, where

$$\mathscr{F}(\tilde{\epsilon}) = \{\varrho \in \mathbb{C} : \|\varrho + \tilde{\beta}\|_\omega \leq 1/(2|\alpha|) + \tilde{\epsilon}\}. \tag{5.12}$$

Thus, the results of Chapter 4 are immediately applicable and even give a prime number theorem of sorts, in the sense that

$$\#\{\text{prime elements } p \in \mathcal{B}_\omega(\alpha, \beta) : N(p) \leq x_k\} \sim \frac{\delta_\mathcal{O}}{|\alpha|^2} \frac{x_k}{\log x_k}$$

along some sequence $(x_k)_k$ tending to infinity, where $\delta_\mathcal{O}$ is given by (A.8).

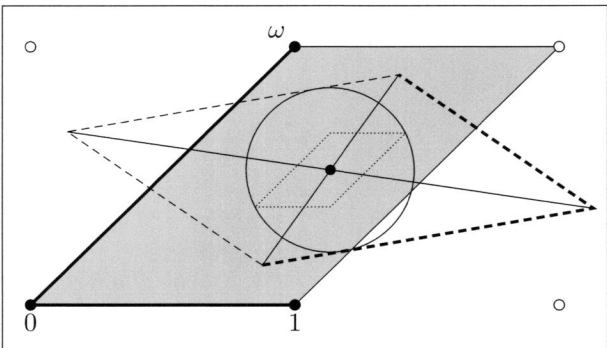

Figure 8: The fundamental parallelogram Λ (grey) associated with ω and $\mathscr{F} = \alpha^{-1}(\beta - \Lambda)$ (dashed). One can fit a circle of diameter $|\sin \arg \omega|$ inside $\mathscr{F}$ and, hence, an *unrotated* rescaled version of Λ (dotted) fitting inside the circle certainly also fits inside $\mathscr{F}$ (disregarding points on the boundary for the moment; the part of the boundary which is contained in the corresponding region is indicated with a thick border).

However, if α (still assumed to exceed 1 in absolute value) has is not a negative real number, then squeezing $\mathscr{F} \bmod \mathcal{O}$ in-between two sets of the shape (5.12) as we did in (5.11) may not be possible. Nonetheless, this problem can be remedied without much additional effort, because one can still exhibit some $\tilde{\beta}$ and some $\delta > 0$ such that

$$\{\varrho \in \mathbb{C} : \|\varrho + \tilde{\beta}\|_\omega \le \delta - \tilde{\epsilon}\} \subset (\mathscr{F} \bmod \mathcal{O}) \tag{5.13}$$

for every $\tilde{\epsilon} > 0$. In order to detect a large proportion of primes, it is desirable to take δ as large as possible. In this regard, by some elementary geometric considerations (compare Fig. 8), one may arrive at the conclusion that for

$$\delta = \frac{1}{2|\alpha|} \cdot \begin{cases} 1 & \text{if } \alpha \in \mathbb{R} \text{ and } \alpha < 0, \\ |\sin \arg \omega| (\operatorname{diam} \Lambda)^{-1} & \text{otherwise} \end{cases} \tag{5.14}$$

there is still some $\tilde{\beta}$ satisfying (5.13). Thus, we may cast the outcome of this section as follows:

Proposition 5.11. *Let* $\mathcal{B}_\omega(\alpha, \beta)$ *be a Beatty-type set as in* (1.10) *with* $|\alpha| \ge 1$. *Then there exists a shift* $\tilde{\beta} = \tilde{\beta}_1 + \tilde{\beta}_2 \omega$ *such that the set*

$$\mathscr{B}(\tilde{\epsilon}) = \{m \in \mathcal{O} : \|m/\alpha + \tilde{\beta}\|_\omega \le \delta - \tilde{\epsilon}\}$$

is contained in $\mathcal{B}_\omega(\alpha, \beta)$ *for every* $\tilde{\epsilon} > 0$.

Remark 5.12. *The above choice of* δ *is not necessarily as large as possible, as can already be seen in Fig. 8 (the dotted parallelogram could be enlarged without crossing the boundary of the dashed rectangle). However, it has the pleasant feature of being of the shape*

$$\delta = \{\text{largest } \delta\text{-value one could possibly hope for, i.e., } 1/(2|\alpha|)\} \times$$
$$\times \{\text{factor } \in (0, 1] \text{ only depending on } \omega, \text{ but not on } \alpha\}.$$

5.4   Proof of the main results

To give a *proof of Theorem 5.5,* it suffices to merely invoke Proposition 5.11 and Corollary 4.6 when $|\alpha| > 1$. On the other hand, if $|\alpha| \le 1$, then pick some integer a such that $|a\alpha| > 1$ and apply the already established part of the result to $\mathcal{B}_\omega(a\alpha, \beta) \subseteq \mathcal{B}_\omega(\alpha, \beta)$. $\square$

It remains to prove Theorem 5.7. We apply Theorem 4.4 with $\vartheta = \alpha^{-1}$, shift $\tilde{\beta}$ obtained from Proposition 5.11, and a and q equal to some $a^{(r)}$ and $q^{(r)}$ obtained from the Hurwitz continued fraction expansion to ϑ, where the particular choice of r is be given later. By Proposition 5.6, we may then take $C = 2 + \sqrt{2}$ in Theorem 4.4. As in the aforementioned theorem we shall write $x = |q|^{12}$ and we suppose from the outset that x be larger than 2^{100} (say). Finally, also taking $\epsilon = \frac{1}{384}$ and writing $M = M_{\mathbb{Z}[\omega]}(\frac{1}{384})$, we find that

$$\frac{1}{4(\delta - \tilde{\epsilon})^2} \sum_{\substack{x/2 \le N(p) < x \\ \|p\vartheta + \tilde{\beta}\|_\omega \le \delta - \tilde{\epsilon}}} 1 > \sum_{x/2 \le N(p) < x} 1 - 2^{136} |\omega|^{14} M^{5/2} x^{1 - 1/384}$$

for every $\tilde{\epsilon} > 0$ provided that $\delta - \tilde{\epsilon} \geq x^{-1/96}$. Hence, on letting $\tilde{\epsilon} \searrow 0$,

$$\frac{1}{4\delta^2} \sum_{\substack{x/2 \leq N(p) < x \\ \|p\vartheta + \tilde{\beta}\|_\omega < \delta}} 1 > \sum_{x/2 \leq N(p) < x} 1 - 2^{136}|\omega|^{14} M^{5/2} x^{1-1/384}$$

if $\delta > x^{-1/96}$. Clearly, if the left hand side in the above inequality is positive, then

$$\bigcup_{\tilde{\epsilon} > 0} \mathscr{B}(\tilde{\epsilon}) \subseteq \mathcal{B}_\omega(\alpha, \beta)$$

must contain a prime element with norm below x. Employing Proposition A.7, we find that the right hand side of the above inequality exceeds

$$\left(1.9 \frac{x^{1/384}}{\log x} - 2^{136}|\omega|^{14} M^{5/2}\right) x^{1-1/384}$$

and this is certainly positive if

$$2^{101} x^{1/408} \geq 2^{136}|\omega|^{14} M^{5/2}.$$

On writing $x = |q|^{12}$, the above inequality may be stated equivalently as

$$|q| \geq 2^{1190}|\omega|^{476} M^{85}. \tag{5.15}$$

Consequently, we find a prime element p with the aforementioned properties if we manage to satisfy (5.15) and $|q| > \delta^{-8}$. To do this, we use the following analogue of Lemma 3.6:

Lemma 5.13. *Let ϑ, R and $q^{(r)}$ ($r \leq R$) be as above. Then*

$$|q^{(r)}| \geq 1.5^{\lfloor r/2 \rfloor}.$$

Proof. In [42, Theorem 5.2] it is proved that $q^{(r)} \geq 1.5\, q^{(r-2)}$ for $r \geq 2$. Therein it is assumed that ϑ belongs to the set given in (5.5), but evidently this is unimportant for the size of the denominator $q^{(r)}$. $\qquad\square$

Therefore,

- $x \geq 2^{100}$ is certainly satisfied if $r \geq 2^5$,

- (5.15) is satisfied if $r \geq 2^{27} \log(2|\omega|) \log M$,

- and $|q| > \delta^{-8}$ is satisfied if $r \geq 2^6 \log(\delta^{-1})$.

This concludes the proof of Theorem 5.7.

Remark 5.14. *Hurwitz [43] also introduced a continued fraction algorithm for the Eisenstein integers. Given the leisure to work out an Eisenstein-analogue of Lemma 5.13, one could write down a corresponding version of Theorem 5.7 for the Eisenstein integers.*

Appendix A

Facts about quadratic extensions

In this chapter, we collect some basic facts concerning the ring of integers $\mathcal{O}$ of an imaginary quadratic number field $\mathbb{K} \subseteq \mathbb{C}$. Proofs of the results we do not prove are readily found in textbooks such as, for instance, [44, 66, 31, 3, 22].

We assume that the reader is familiar with the theorem about unique factorisation of non-trivial ideals $\mathfrak{a} \subset \mathcal{O}$ into prime ideals (see, e.g., [66, I, §3] or [3, Chapter 5] for a broader context), our discussion of class number 1 in Section 1.2.3 and the notation from Section 4.1.2.

A.1 The choice of the generator ω

Since our work in Chapters 4 and 5 depends not only on $\mathcal{O}$ but also on the choice of the generator ω, we deem it worthwhile to comment briefly on the shape of admissible values for ω. Every imaginary quadratic number field $\mathbb{K} \subseteq \mathbb{C}$ is of the form $\mathbb{K} = \mathbb{Q}(\sqrt{d})$ with some negative square-free integer d. Then the *discriminant* D of $\mathbb{K}$ is given by

$$D = \begin{cases} 4d & \text{if } d \not\equiv 1 \bmod 4, \\ d & \text{if } d \equiv 1 \bmod 4, \end{cases} \tag{A.1}$$

and it satisfies $\mathbb{K} = \mathbb{Q}(\sqrt{D})$ and

$$\mathcal{O} = \mathbb{Z}[\tilde{\omega}], \quad \text{where } \tilde{\omega} = \frac{1}{2}(D + \sqrt{D}). \tag{A.2}$$

It is a trivial matter to check by a simple calculation or by appealing to Lemma A.2 below that, for any ω such that $\mathcal{O} = \mathbb{Z}[\omega]$, one has $\Im\omega = \pm\Im\tilde{\omega}$ and, consequently,

$$\omega = k \pm \tilde{\omega} \tag{A.3}$$

for some integer k and some choice of sign. On the other hand, the right hand side of (A.3) always yields a value ω with $\mathcal{O} = \mathbb{Z}[\omega]$.

A.2 The norm of an element

Recall the definition of the norm $N(m)$ of an element $m \in \mathcal{O}$ from Section 4.1.2. This norm is always a non-negative integer and vanishes if and only if $m = 0$. The elements with norm 1 are precisely the units of $\mathcal{O}$ and they are determined by the following proposition:

Proposition A.1. *The units of $\mathcal{O}$ are given by*

- $\pm 1, \pm i$ *if $\mathbb{K} = \mathbb{Q}(i)$,*

- $\pm 1, \pm \zeta, \pm \zeta^2$ *if $\mathbb{K} = \mathbb{Q}(\sqrt{-3})$ (here $\zeta = \exp(2\pi i/6)$),*

- ± 1 *otherwise.*

In particular, $\mathcal{O}$ contains at most 6 units.

The last part of the above proposition is used throughout this thesis without explicit reference to it.

Furthermore, we need some information about the number of elements of $\mathcal{O}$ with norm below some quantity. To this end, note that $N(n_1 + n_2\omega)$ is a positive-definite quadratic form in the real variables n_1 and n_2 and that there is a general theory providing estimates for

$$\#\{n \in \mathcal{O} : N(n) \le y\}$$

(see, e.g., the monograph of Krätzel [52]). Here, however, even the following rough estimate is sufficient:

Lemma A.2. *We have*

$$\#\{n \in \mathcal{O} : N(n) \le y\} = \frac{\pi y}{|\Im\omega|}(1 + o(1))$$

as $y \to \infty$, where the implied constant only depends on $|\Im\omega|$.

As already noted in Appendix A.1, this shows that $|\Im\omega|$ is independent of the particular choice of a generator ω of $\mathcal{O}$. From (A.2), we then conclude that

$$|\Im\omega| \ge \frac{\sqrt{3}}{2}. \tag{A.4}$$

Furthermore, adapting [52, pp. 16–17] to this setting and using the above bound, one readily establishes the following explicit bound which we (often tacitly) use throughout.

Lemma A.3. *For $y \ge \frac{1}{2}$, $\#\{n \in \mathcal{O} : N(n) \le y\} \le 13y$.*

A.3   Primes in $\mathcal{O}$

In the sieve-framework presented in Appendix B one is essentially sifting by prime ideals, but most of what we do is stated in terms of elements of $\mathcal{O}$, where our restriction to class number 1 allows one to identify ideals with elements up to multiplication by units. In this regard it is convenient to introduce a set of prime elements of $\mathcal{O}$ where associates have been discarded. Just for the sake of having an unambiguous definition of this, note that for each prime element $\tilde{p} \in \mathcal{O}$ there is an associated prime element p with minimal argument $\arg p \in [0, 2\pi)$. Let $\mathbb{P}_{\mathcal{O}}$ be the set of those p just mentioned and put

$$\mathbb{P}_{\mathcal{O}}(z) = \{p \in \mathbb{P}_{\mathcal{O}} : N(p) < z\}. \tag{A.5}$$

A.3.1 Ramification

We shall describe the prime ideals in $\mathcal{O}$ in a little more detail. (Here one does not have to restrict to class number 1.) To this end, note that for each prime ideal $\mathfrak{p} \subseteq \mathcal{O}$ its restriction to $\mathbb{Z}$, namely $\mathfrak{p} \cap \mathbb{Z}$, is a prime ideal in $\mathbb{Z}$ and, thus, of the form (p) for some rational prime p. For the norm $N\mathfrak{p} = \#(\mathcal{O}/\mathfrak{p})$ we have $N\mathfrak{p} \in \{p, p^2\}$ and we say that $\mathfrak{p}$ is *lying over* p.

From ramification theory (see, e.g., [66, I, §8]) one knows that the number k_p of distinct prime ideals $\mathfrak{p} \subseteq \mathcal{O}$ lying over a given rational prime p is bounded by $\dim_{\mathbb{Q}} \mathbb{K} = 2$ and greater than 0. For every rational prime p, one and only one of the following three cases occurs:

- $k_p = 2$ and the two distinct prime ideals $\mathfrak{p}, \tilde{\mathfrak{p}}$ lying over p each have norm p.

- $k_p = 1$ and the prime ideal $\mathfrak{p}$ lying over p satisfies $\mathfrak{p}^2 = p\mathcal{O}$ and has norm p.

- $k_p = 1$ and the prime ideal $\mathfrak{p}$ lying over p is given by $p\mathcal{O}$ and has norm p^2.

We only mention in passing that the discriminant D defined in (A.1) regulates which of the above cases applies to an individual value of p (see, e.g., [44, Propositions 13.1.3 and 13.1.4]).

A.3.2 Ordering the primes

Let $\mathcal{O}$ have class number 1 and consider the set $\mathbb{P}_{\mathcal{O}}$ as defined in the beginning of Appendix A.3. We introduce a strong total order on $\mathbb{P}_{\mathcal{O}}$ as follows:

$$q \prec p \iff (p \succ q \iff) \begin{cases} N(q) < N(p), \text{ or} \\ N(q) = N(p) \text{ and } \arg q < \arg p. \end{cases} \tag{A.6}$$

Let

$$\Pi(p) = \prod_{q \prec p} q. \tag{A.7}$$

The presence of the case $N(q) = N(p)$ in (A.6) requires some additional work in Appendix B. However, recalling our restriction to class number 1 and appealing to the discussion of possible ramification behaviour in Appendix A.3.1, we infer that, for any prime element p of $\mathcal{O}$, there is *at most one* prime element q satisfying both

$$q \prec p \quad \text{and} \quad N(q) = N(p).$$

Example A.4 (Eisenstein integers). *Consider the imaginary quadratic number field $\mathbb{K} = \mathbb{Q}(\sqrt{-3})$ with its ring of integers $\mathcal{O}$, the Eisenstein integers. Fig. 9 shows elements of $\mathcal{O}$ with small norm. Each of the six sectors sketched therein contains a set of pairwise non-associate Eisenstein integers which, together with their associates, yield the whole set $\mathcal{O} \setminus \{0\}$ (assuming the sectors to be half-open and excluding zero from the discussion). Denote the sector containing the solid arcs by $\mathscr{S}$ and assume that $\mathscr{S}$ contains the positive reals (thick line) but no other of its boundary points. Then the prime Eisenstein integers contained in $\mathscr{S}$ are precisely those contained in $\mathbb{P}_{\mathcal{O}}$. As an illustration of our previous discussion of primes $q \prec p$ with equal norm, one may note that the solid arcs contained in $\mathscr{S}$ only ever intersect at most two filled in points.*

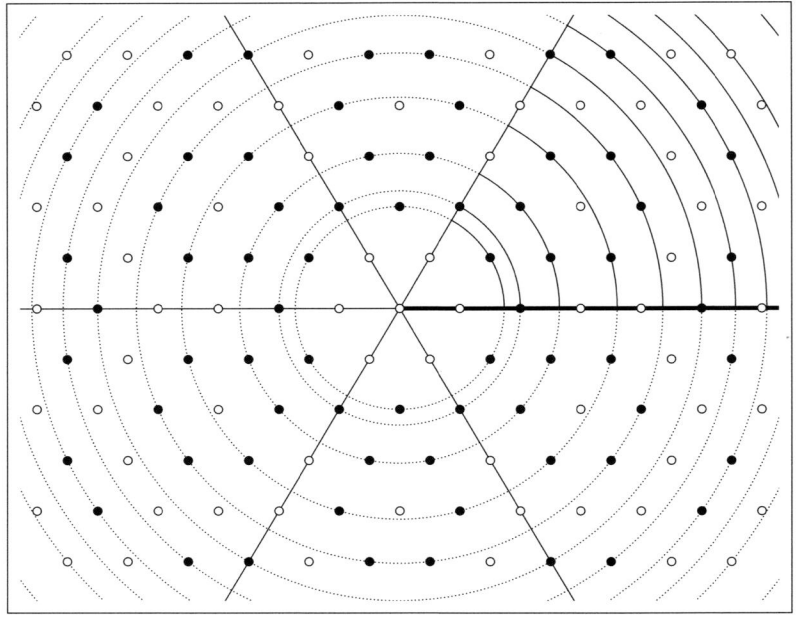

Figure 9: Eisenstein integers $\mathcal{O} = \mathbb{Z}[\omega]$ ($\omega = e(1/3)$) with small norm (the points). The points corresponding to prime elements p are filled in and circles with radii equal to $\sqrt{N(p)}$ are sketched. The symmetry with respect to the sectors is due to the unit group of $\mathcal{O}$ consisting of all sixth roots of unity.

A.3.3 Prime (ideal) distribution

We need some information about the number of prime elements in rings of integers $\mathcal{O}$ of imaginary quadratic number fields with class number 1. To this end, we note the following special case of Landau's prime ideal theorem [56]:

Theorem A.5. *Let $\mathcal{O}$ be the ring of integers of an imaginary quadratic number field $\mathbb{K}$ (not necessarily with class number 1). Then*

$$\#\{\textit{prime ideals } \mathfrak{p} \subseteq \mathcal{O} : N\mathfrak{p} \leq x\} = \frac{x}{\log x}(1 + o_{\mathcal{O}}(1)) \quad \textit{as } x \to \infty.$$

As an immediate consequence (mind the discussion in Appendix A.3.1), we may state the following:

Corollary A.6. *Suppose that $\mathbb{K}$ has class number 1. Then*

$$\#\{\textit{prime elements } p \in \mathcal{O} : N(p) \leq x\} \asymp \frac{x}{\log x} \quad \textit{as } x \to \infty.$$

Of course, the above corollary can easily be sharpened to an asymptotic formula

$$\#\{\text{prime elements } p \in \mathcal{O} : N(p) \leq x\} = \frac{x}{\log x}(\delta_{\mathcal{O}} + o(1)), \qquad (A.8)$$

by a closer analysis of the splitting behaviour of prime ideals $(p) \subseteq \mathbb{Z}$ as hinted upon in Appendix A.3.1, supplemented by the prime number theorem for arithmetic progressions, but we does not need this here. However, for our work in Chapter 5 we require some explicit version of this at least for the Gaussian integers. For $x \geq 2$, using [31, Theorem 252], one immediately finds that

$$\#\{\text{prime elements } p \in \mathbb{Z}[i] : N(p) \leq x\}$$
$$= \#\{\text{units in } \mathbb{Z}[i]\} \cdot (1 + 2\pi(x, 1 \bmod 4) + \pi(\sqrt{x}, 3 \bmod 4)),$$

where

$$\pi(x, a \bmod m) = \#\{\text{rational primes } p \leq x : p \equiv a \bmod m\}.$$

Using this together with results of Ramaré and Rumely [68], a standard partial summation and some numerical calculations, it is an easy matter to establish the following Chebyshev-type prime number theorem:

Proposition A.7. *For* $x \geq 2^{65}$,

$$\#\{\text{prime elements } p \in \mathbb{Z}[i] : N(p) < x\} = \frac{x}{\log x}(4 + O^*(0.0298)).$$

A.4 Divisor functions in $\mathcal{O}$

Let $d_k(\mathfrak{a})$ denote the number of representations of the ideal $\mathfrak{a} \subseteq \mathcal{O}$ as a product of precisely k ideals and let $d(\mathfrak{a}) = d_2(\mathfrak{a})$. Just as its $\mathbb{Z}$-analogue, this function is multiplicative and on powers of prime ideals it is given by

$$d_k(\mathfrak{p}^\nu) = \binom{\nu + k - 1}{k - 1}. \tag{A.9}$$

In Chapter 4 and Appendix B we require some bounds for $d(\mathfrak{a})$ and averages of d and d_4. Of course, results of these types are well-known even in a more general semi-group setting (see, e.g., [51, Ch. 4, §4]), but the implied constants are seldom computed. Thus, we permit ourselves to explicitly adapt some of the standard proofs, computing the implied constants along the way.

Lemma A.8 (The divisor bound). *For any* $\epsilon > 0$ *there is some* $M_\mathcal{O}(\epsilon) \geq 1$ *such that for any ideal* $\mathfrak{a} \subseteq \mathcal{O}$,

$$d(\mathfrak{a}) \leq M_\mathcal{O}(\epsilon)(N\mathfrak{a})^\epsilon.$$

Moreover, one may take

$$M_\mathcal{O}(\epsilon) = \prod_{N\mathfrak{p} \leq e^{1/\epsilon}} \max_{\nu \in \mathbb{N}_0} \frac{\nu + 1}{(N\mathfrak{p}^\nu)^\epsilon}.$$

Proof. Let $\mathfrak{a} = \mathfrak{p}_1^{\nu_1} \cdots \mathfrak{p}_r^{\nu_r}$ denote the factorisation of $\mathfrak{a}$ into prime ideals. Then

$$\frac{d(\mathfrak{a})}{(N\mathfrak{a})^\epsilon} = \prod_{k=1}^r \frac{\nu_k + 1}{(N\mathfrak{p}_k^{\nu_k})^\epsilon}.$$

If $N\mathfrak{p}_k \geq e^{1/\epsilon}$, then $(N\mathfrak{p}_k^{\nu_k})^\epsilon \geq 1 + \nu_k$. Therefore, we have

$$\frac{d(\mathfrak{a})}{(N\mathfrak{a})^\epsilon} \leq \prod_{N\mathfrak{p} \leq e^{1/\epsilon}} \max_{\nu \in \mathbb{N}_0} \frac{\nu + 1}{(N\mathfrak{p}^\nu)^\epsilon},$$

as required. $\qquad\qquad\square$

The next result is quite crude, but for our purposes it suffices.

Lemma A.9. *For* $x \geq 2$,

- $\displaystyle\sum_{N\mathfrak{a} \leq x} d(\mathfrak{a}) < 2^7 x (\log x)^4,$

- $\displaystyle\sum_{N\mathfrak{a} \leq x} d(\mathfrak{a})^2 < 2^{10} x (\log x)^8,$

- $\displaystyle\sum_{N\mathfrak{a} \leq x} d_4(\mathfrak{a}) < 2^{13} x (\log x)^8.$

The proof we give below (compare [45, p. 23]) rests on a version of Rankin's trick and requires the following Mertens-type estimates (again, the exponent of the logarithm here is not optimal):

Lemma A.10. *For* $x \geq 2$,

$$\prod_{N\mathfrak{p} \leq x} (1 + (N\mathfrak{p})^{-1}) < 2.25 (\log x)^2, \qquad\qquad (\text{A.}10)$$

$$\prod_{N\mathfrak{p} \leq x} (1 + (N\mathfrak{p})^{-2}) < 2.5. \qquad\qquad (\text{A.}11)$$

Proof. Upon taking the logarithm of the products in question, we find that

$$L_k = \log \prod_{N\mathfrak{p} \leq x} (1 + (N\mathfrak{p})^{-k}) = \sum_{N\mathfrak{p} \leq x} \log(1 + (N\mathfrak{p})^{-k}) \leq \sum_{N\mathfrak{p} \leq x} (N\mathfrak{p})^{-k}.$$

From Appendix A.3.1 we know that $N\mathfrak{p}$ is either the square of a rational prime or itself a rational prime; furthermore, given some rational prime p, there are at most two distinct prime ideals $\mathfrak{p}$ with $p \mid N\mathfrak{p}$. Thus,

$$L_k \leq 2 \sum_{p \leq x} p^{-k}.$$

In particular, for any $y > 0$,

$$L_2 < 2 \sum_{p < y} p^{-2} + \sum_{p \geq y} \frac{2}{(p-1)p} < 2 \sum_{p < y} p^{-2} + \frac{2}{(y-1)}.$$

Taking, e.g., $y = 40$, we deduce (A.11). Moreover, the claimed bound (A.10) is obtained upon noting that [70, (3.20)] gives

$$\exp\left(\sum_{p \leq x} p^{-1}\right) < \frac{3}{2}\log x. \qquad \square$$

Proof of Lemma A.9. Let f be a complex-valued multiplicative function on ideals of $\mathcal{O}$. Then we have

$$\sum_{N\mathfrak{a} \leq x} f(\mathfrak{a}) \leq x \sum_{N\mathfrak{a} \leq x} \frac{f(\mathfrak{a})}{N\mathfrak{a}} \leq x \prod_{N\mathfrak{p} \leq x} \sigma_{\mathfrak{p}}(f), \tag{A.12}$$

where

$$\sigma_{\mathfrak{p}}(f) = \sum_{\nu=0}^{\infty} \frac{f(\mathfrak{p}^{\nu})}{N\mathfrak{p}^{\nu}}$$

(see [45, p. 23]). To prove the assertions of the lemma, note that, using (A.9), one easily verifies that

$$\sigma_{\mathfrak{p}}(d) = \sum_{\nu=0}^{\infty} \frac{\nu+1}{N\mathfrak{p}^{\nu}} = \frac{(N\mathfrak{p})^2}{(N\mathfrak{p}-1)^2} \leq (1+(N\mathfrak{p})^{-1})^2(1+(N\mathfrak{p})^{-2})^3,$$

$$\sigma_{\mathfrak{p}}(d^2) = \sum_{\nu=0}^{\infty} \frac{(\nu+1)^2}{N\mathfrak{p}^{\nu}} = \frac{(N\mathfrak{p})^2(N\mathfrak{p}+1)}{(N\mathfrak{p}-1)^3} \leq (1+(N\mathfrak{p})^{-1})^4(1+(N\mathfrak{p})^{-2})^4$$

and

$$\sigma_{\mathfrak{p}}(d_4) = \sum_{\nu=0}^{\infty} \frac{1}{N\mathfrak{p}^{\nu}}\binom{\nu+3}{3} = \frac{(N\mathfrak{p})^4}{(N\mathfrak{p}-1)^4} \leq (1+(N\mathfrak{p})^{-1})^4(1+(N\mathfrak{p})^{-2})^6.$$

On plugging these inequalities into (A.12) and invoking Lemma A.10, one obtains the asserted statements. $\qquad \square$

Appendix B

Harman's sieve method

In this chapter we give a proof of Harman's sieve method as formulated in Theorem 4.7. Our exposition closely follows Harman [35, Chapter 3], who proves his sieve result for $\mathbb{Z}$. Practically the same proof also works for $\mathcal{O}$, but a minor technical complication occurs in Appendix B.2.3 due to the possible existence of non-associate primes in $\mathcal{O}$ with equal norm (see Eqs. (B.13) and (B.14) below and the arguments that follow them).

We add that there is another noteworthy point where we differ from [35]: the coefficients b_n in Theorem 4.7 are assumed to be divisor-bounded, but [35, Theorem 3.1] supposedly only needs to supply the analogue of (4.9) with $|b_n| \leq 1$. However, we do not see how this follows from the proof given in [35], which in fact turns out to be almost verbatim the same as in [34, Lemma 2], although there, b_n is assumed to be divisor-bounded. We asked Harman via email if he cared to clarify the matter, but as of the writing of this thesis, we did not receive a response.

B.1   A variant of Perron's formula

In the course of the proof of Theorem 4.7 we shall need to disentangle summation variables joined by some inequality. A well-known standard technique for achieving this is furnished by the following lemma:

Lemma B.1. *Suppose that one is given distinct real numbers $\rho, \gamma > 0$ and $T \geq 1$ is some real parameter. Then*

$$\left| \mathbf{1}_{\{\gamma < \rho\}} - \frac{1}{\pi} \int_{-T}^{T} e^{i\gamma t} \frac{\sin(\rho t)}{t}\, \mathrm{d}t \right| \leq \frac{8}{\pi T |\gamma - \rho|}. \tag{B.1}$$

This is being applied in full detail in Appendix B.2.3, but the lengthy formulae therein somewhat shroud the picture of what is actually going on. Therefore, we deem it worthwhile to give two examples illustrating the point to be made:

Example B.2. *Consider, for instance, a finite double sum of the type*

$$\sum \sum_{m < n} a_m b_n \quad (m, n \in \mathbb{N}),$$

which trivially can be rewritten as

$$\sum \sum_{m,n} \mathbf{1}_{\{m<n\}} a_m b_n$$

with independent summation variables, but here the new summation expression, i.e., $\mathbf{1}_{\{m<n\}} a_m b_n$*, does not factor into a product* $\tilde{a}_m \tilde{b}_n$ *(say). However, if one replaces*

$$\mathbf{1}_{\{m<n\}} = \mathbf{1}_{\{m+\frac{1}{2}<n\}} \tag{B.2}$$

by the integral from the above lemma, then the resulting expression

$$\frac{1}{\pi} \int_{-T}^{T} \sum_{m,n} \sum (a_m e^{imt} e^{\frac{it}{2}})(b_n \sin(nt)) \frac{dt}{t}$$

involves an integral of a double sum in which the summed expression factors are as desired. (Note that (B.2) holds since we only apply this for integers m, n*, and this step is required in order not to run into trouble with the term* $|\gamma - \rho|^{-1}$ *in (B.1), when applying the lemma for terms with* $m = n$*.)*

Example B.3. *Similarly, one may deal with finite sums of the type*

$$\sum_{mn<x} \sum a_m b_n \quad (1 \neq m, n, x \in \mathbb{N}),$$

but naively putting $\gamma = mn$ *and* $\rho = x$ *in the above lemma is of no use, because the resulting factor* $e^{i\gamma t} = e^{imnt}$ *does not split in the desired way. To get around this, one may take the logarithm on both sides of* $mn < x - \frac{1}{2}$*; then, for* $\gamma = \log mn$*, we have a decomposition of the desired type:* $e^{i\gamma t} = m^{it} n^{it}$*.*

Proof of Lemma B.1. This is a standard exercise in contour integration and may be found, for instance, in [23, Lemma 5.5.1]. Since there the right hand side of (B.1) is only given as $\ll (T|\gamma - \rho|)^{-1}$, we add some extra details. Indeed, from [23, (5.57)] we know that, for any non-zero $\lambda \in \mathbb{R}$,

$$\left| \lim_{\epsilon \searrow 0} \left\{ \int_{-T}^{-\epsilon} + \int_{\epsilon}^{T} \right\} \frac{e^{i\lambda t}}{it} \, dt - \pi \operatorname{sgn} \lambda \right| \leq \frac{4}{T|\lambda|}.$$

Therefore, we have

$$\int_{-T}^{T} e^{i\gamma t} \frac{\sin(\rho t)}{t} \, dt = \frac{1}{2} \int_{-T}^{T} \frac{e^{i(\gamma+\rho)t} - e^{i(\gamma-\rho)t}}{it} \, dt$$

$$= \frac{1}{2} \sum_{\theta=\pm 1} \theta \lim_{\epsilon \searrow 0} \left\{ \int_{-T}^{-\epsilon} + \int_{\epsilon}^{T} \right\} \frac{e^{i(\gamma+\theta\rho)t}}{it} \, dt$$

$$= \frac{\pi}{2} \operatorname{sgn}(\gamma + \rho) - \frac{\pi}{2} \operatorname{sgn}(\gamma - \rho) + O^*\left(\frac{8}{T|\gamma - \rho|}\right)$$

$$= \pi \cdot \mathbf{1}_{\{\gamma<\rho\}} + O^*\left(\frac{8}{T|\gamma - \rho|}\right).$$

The assertion of the lemma now follows after dividing both sides by π. □

B.2 Proof of the sieve result

We move on to the actual proof of Theorem 4.7.

B.2.1 The difference $S(\mathscr{A}, z) - \lambda S(\mathscr{B}, z)$

For a non-unit d, let $\mu(d)$ be defined as $(-1)^r$ if d is the product of precisely r non-associate prime elements and $\mu(d) = 0$ otherwise. If d is a unit, then put $\mu(d) = 1$. Write $z = x^\kappa$ and recall the definition (A.5) of $\mathbb{P}_\mathcal{O}(z)$. Now, writing

$$P_\mathcal{O}(z) = \prod_{p \in \mathbb{P}_\mathcal{O}(z)} p,$$

for $\mathscr{X} \subseteq \mathscr{B}$,

$$S(\mathscr{X}, z) = \sum_{\substack{r \in \mathscr{X} \\ r \text{ coprime to } P_\mathcal{O}(z)}} 1 = \sum_{r \in \mathscr{X}} \sum_{\substack{m \mid P_\mathcal{O}(z) \\ m \mid r}} \mu(m)$$

$$= \sum_{m \mid P_\mathcal{O}(z)} \sum_{\substack{r \in \mathscr{X} \\ m \mid r}} \mu(m) = \sum_{m \mid P_\mathcal{O}(z)} \mu(m) \sum_{nm \in \mathscr{X}} 1.$$

Write

$$\Delta(m) = \sum_{nm \in \mathscr{A}} 1 - \lambda \sum_{nm \in \mathscr{B}} 1. \tag{B.3}$$

Then

$$S(\mathscr{A}, z) - \lambda S(\mathscr{B}, z) = \sum_{m \mid P_\mathcal{O}(z)} \mu(m)\Delta(m). \tag{B.4}$$

The sum on the right hand side of (B.4) can be decomposed according to the size of $N(m)$:

$$\left\{ \sum_{\substack{m \mid P_\mathcal{O}(z) \\ N(m) < M}} + \sum_{\substack{m \mid P_\mathcal{O}(z) \\ N(m) \geq M}} \right\} \mu(m)\Delta(m) = S_{\mathrm{I}} + S_{\mathrm{II}}, \quad \text{say.} \tag{B.5}$$

Here

$$S_{\mathrm{I}} = \sum_{\substack{m \mid P_\mathcal{O}(z) \\ N(m) < M}} \mu(m)\Delta(m) = \sum_{\substack{mn \in \mathscr{A} \\ N(m) < M}} \sum a_m - \lambda \sum_{\substack{mn \in \mathscr{B} \\ N(m) < M}} \sum a_m,$$

with

$$a_m = \begin{cases} \mu(m) & \text{if } m \mid P_\mathcal{O}(z), \\ 0 & \text{otherwise.} \end{cases}$$

Hence, by (4.8),

$$|S_{\mathrm{I}}| \leq Y_{\mathrm{I}}. \tag{B.6}$$

Since $\Delta(m) = 0$ when $N(m) \geq x$, S_{II} vanishes if $M \geq x$. Hence, in all that follows, we shall assume $M < x$.

B.2.2 Arranging sums with variables in the correct ranges

Let $g : \mathcal{O} \to \mathbb{C}$ be any function with $g(m) = g(\tilde{m})$ whenever m and $\tilde{m}$ are associates. We group the terms of the sum

$$S = \sum_{m|P_{\mathcal{O}}(z)} \mu(m)g(m)$$

according to the largest prime factor p_1 of m (w.r.t. $\prec$):

$$S = g(1) - \sum_{p_1 \in P_{\mathcal{O}}(z)} \sum_{d|\Pi(p_1)} \mu(d)g(p_1 d), \tag{B.7}$$

where Π is given by (A.7). Evidently, the process giving (B.7) also works if $P_{\mathcal{O}}(z)$ is replaced by $\Pi(p)$; for any $r \in \mathcal{O}$ one has

$$\sum_{d|\Pi(p_1)} \mu(d)g(rd) = g(r) - \sum_{p_2 \prec p_1} \sum_{d|\Pi(p_2)} \mu(d)g(rp_2 d), \tag{B.8}$$

where we have used the relation $\prec$ defined in (A.6).

Let

$$\mathbb{P}_{\mathcal{O}}(z) = \{p_1 \in \mathbb{P}_{\mathcal{O}}(z) : N(p_1) > x^\mu\} \cup \{p_1 \in \mathbb{P}_{\mathcal{O}}(z) : N(p_1) \leq x^\mu\}$$
$$= \mathscr{P}_1 \cup \mathscr{Q}_1, \quad \text{say,}$$

and inductively for $s = 2, 3, \ldots,$

$$\mathscr{Q}'_s = \{(p_1, \ldots, p_{s-1}, p_s) \in (\mathbb{P}_{\mathcal{O}}(z))^s : p_s \prec p_{s-1}, (p_1, \ldots, p_{s-1}) \in \mathscr{Q}_{s-1}\}$$
$$= \mathscr{P}_s \cup \mathscr{Q}_s,$$

where

$$\mathscr{P}_s = \{(p_1, \ldots, p_{s-1}, p_s) \in \mathscr{Q}'_s : N(p_1 \cdots p_{s-1} p_s) > x^\mu\},$$
$$\mathscr{Q}_s = \{(p_1, \ldots, p_{s-1}, p_s) \in \mathscr{Q}'_s : N(p_1 \cdots p_{s-1} p_s) \leq x^\mu\}.$$

Assuming that g vanishes on arguments r with $N(r) \leq x^\mu$, and on applying (B.7) and (B.8),

$$S = -\left\{ \sum_{p_1 \in \mathscr{P}_1} + \sum_{p_1 \in \mathscr{Q}_1} \right\} \sum_{d|\Pi(p_1)} \mu(d)g(p_1 d)$$
$$= - \sum_{p_1 \in \mathscr{P}_1} \sum_{d|\Pi(p_1)} \mu(d)g(p_1 d) + \sum_{(p_1,p_2) \in \mathscr{P}_2} \sum_{d|\Pi(p_2)} \mu(d)g(p_1 p_2 d)$$
$$+ \sum_{(p_1,p_2) \in \mathscr{Q}_2} \sum_{d|\Pi(p_2)} \mu(d)g(p_1 p_2 d).$$

The identity (B.8) is also applicable to the last sum and it transpires that an induction gives

$$S = \sum_{s \leq t} (-1)^s \sum_{(p_1, p_2, \ldots, p_s) \in \mathscr{P}_s} \sum_{d | \Pi(p_s)} \mu(d) g(p_1 p_2 \cdots p_s d) + $$
$$+ (-1)^t \sum_{(p_1, p_2, \ldots, p_t) \in \mathscr{Q}_t} \sum_{d | \Pi(p_t)} \mu(d) g(p_1 p_2 \cdots p_t d)$$

for any $t \in \mathbb{N}$. Since the product of t primes has norm $\geq 2^t$, we have

$$\mathscr{Q}_t = \emptyset \quad \text{for} \quad t > \frac{\mu}{\log 2} \log x.$$

Hence,

$$S = \sum_{s \leq t} (-1)^s \sum_{(p_1, p_2, \ldots, p_s) \in \mathscr{P}_s} \sum_{d | \Pi(p_s)} \mu(d) g(p_1 p_2 \cdots p_s d)$$

for $t = \lfloor \frac{1}{\log 2} \log x \rfloor + 1$, say.

We apply this to S_{II} from (B.5) with

$$g(m) = \begin{cases} \Delta(m) & \text{if } N(m) \geq M, \\ 0 & \text{otherwise.} \end{cases}$$

Note that, because of $M > x^\mu$, $g(r) = 0$ for all r with $N(r) \leq x^\mu$. Thus,

$$S_{\mathrm{II}} = \sum_{s \leq t} (-1)^s S_{\mathrm{II}}(s), \tag{B.9}$$

where

$$S_{\mathrm{II}}(s) = \sum_{\substack{(p_1, \ldots, p_s) \in \mathscr{P}_s \\ m := p_1 \cdots p_s}} \sum_{\substack{d | \Pi(p_s) \\ N(md) \geq M}} \mu(d) \Delta(md).$$

Another application of (B.8) gives

$$S_{\mathrm{II}}(s) = \sum_{\substack{(p_1, \ldots, p_s) \in \mathscr{P}_s \\ m := p_1 \cdots p_s \\ N(m) \geq M}} \mu(m) \Delta(m) - \sum_{\substack{(p_1, \ldots, p_s) \in \mathscr{P}_s \\ m := p_1 \cdots p_s}} \sum_{p \prec p_s} \sum_{\substack{d | \Pi(p) \\ N(mpd) \geq M}} \mu(d) \Delta(mpd)$$

$$= S_{\mathrm{II},1}(s) - S_{\mathrm{II},2}(s), \quad \text{say.} \tag{B.10}$$

B.2.3 Estimating Type II sums

Given $m = p_1 \cdots p_{s-1} p_s$ with

$$(p_1, \ldots, p_{s-1}, p_s) \in \mathscr{P}_s \quad \text{and} \quad (p_1, \ldots, p_{s-1}) \in \mathscr{Q}_{s-1},$$

and noting that $N(p_s) \leq N(p_1) < z = x^\kappa$, we have

$$x^\mu < N(m) = N(p_1 \cdots p_{s-1}) N(p_s) < x^\mu x^\kappa.$$

Using this, we find that $S_{\text{II},1}(s)$ can be expressed as

$$\sum_{mn\in\mathscr{A}}\sum a_m - \lambda \sum_{mn\in\mathscr{B}}\sum a_m,$$

where the coefficients

$$a_m = \mu(m)\,\mathbf{1}_{\{N(m)\geq M\}}\,\mathbf{1}_{\{p_1\cdots p_s:(p_1,\ldots,p_s)\in\mathscr{P}_s\}}(m)$$

are only supported on m with $x^\mu < N(m) < x^{\mu+\kappa}$. Hence, by (4.9),

$$|S_{\text{II},1}(s)| \leq Y_{\text{II}}. \tag{B.11}$$

Moving on to $S_{\text{II},2}(s)$, we expand the definition (B.3) of Δ, getting

$$S_{\text{II},2}(s) = S_{\text{II},2}(s,\mathscr{A}) - \lambda S_{\text{II},2}(s,\mathscr{B}), \tag{B.12}$$

where

$$S_{\text{II},2}(s,\mathscr{X}) = \sum_{\substack{(p_1,\ldots,p_s)\in\mathscr{P}_s \\ m:=p_1\cdots p_s}} \sum_{p\prec p_s} \sum_{\substack{d\mid\Pi(p) \\ N(mpd)\geq M}} \mu(d) \sum_{\ell mpd\in\mathscr{X}} 1$$

$$= \sum_{\substack{(p_1,\ldots,p_s)\in\mathscr{P}_s \\ m:=p_1\cdots p_s}} \sum_{nm\in\mathscr{X}} \sum_{p\prec p_s} \sum_{\substack{d\mid\Pi(p) \\ \ell pd=n \\ N(mpd)\geq M}} \mu(d).$$

In order to apply (4.9), we must disentangle the variables m and n in the above summation. To this end, split

$$\sum_{p\prec p_s} = \sum_{\substack{p\prec p_s \\ N(p)=N(p_s)}} + \sum_{\substack{p\prec p_s \\ N(p)<N(p_s)}} \tag{B.13}$$

to obtain a decomposition

$$S_{\text{II},2}(s,\mathscr{X}) = S_{\text{II},2}^{=}(s,\mathscr{X}) + S_{\text{II},2}^{<}(s,\mathscr{X}), \quad \text{say.} \tag{B.14}$$

For $S_{\text{II},2}^{<}(s,\mathscr{X})$ we have

$$S_{\text{II},2}^{<}(s,\mathscr{X}) = \sum_{\substack{(p_1,\ldots,p_s)\in\mathscr{P}_s \\ m:=p_1\cdots p_s}} \sum_{nm\in\mathscr{X}} \sum_{p\in\mathbb{P}_\mathcal{O}(z)} \sum_{\substack{d\mid\Pi(p) \\ \ell pd=n}} \mu(d)\chi(m,d,p,p_s),$$

where

$$\chi(m,d,p,p_s) = \mathbf{1}_{\{N(mpd)\geq M\}}\mathbf{1}_{\{N(p)<N(p_s)\}}$$

and the sum $S_{\text{II},2}^{<}(s,\mathscr{X})$ can be expressed similarly, but needs a little more care: as discussed in Appendix A.3.2, the first summation on the right hand side of (B.13) contains at most one term and we shall write $\mathscr{P}_s'$ for the set of $(p_1,\ldots,p_s)\in\mathscr{P}_s$ for which there is such a term, that is, some $p \prec p_s$ with $N(p) = N(p_s)$. Furthermore, let $\mathbb{P}_\mathcal{O}(z)'$ denote the set of the ps just mentioned, i.e.,

$$\mathbb{P}_\mathcal{O}(z)' = \{p \in \mathbb{P}_\mathcal{O}(z) : \exists p_s : p \prec p_s, \ N(p) = N(p_s)\}.$$

Thus,

$$S_{\overline{\overline{\text{II}}},2}(s, \mathscr{X}) = \sum_{\substack{(p_1,\dots,p_s)\in\mathscr{P}'_s \\ m:=p_1\cdots p_s}} \sum_{nm\in\mathscr{X}} \sum_{p\in\mathbb{P}_\mathcal{O}(z)'} \sum_{\substack{d|\Pi(p) \\ \ell p d=n}} \mu(d)\tilde{\chi}(m, d, p, p_s),$$

where

$$\begin{aligned}\tilde{\chi}(m, d, p, p_s) &= \mathbf{1}_{\{N(mpd)\geq M\}}\mathbf{1}_{\{N(p)=N(p_s)\}} \\ &= \mathbf{1}_{\{N(mpd)\geq M\}}\mathbf{1}_{\{N(p)\leq N(p_s)\}} - \chi(m, d, p, p_s).\end{aligned} \qquad (\text{B.15})$$

We pick some real number ϱ with $|\varrho| \leq \frac{1}{2}$ (depending only on M) such that $\{M + \varrho\} = \frac{1}{2}$ and for $m, p, d \in \mathcal{O}$ the condition $N(mpd) \geq M$ is equivalent to $\log N(mpd) \geq \log(M + \varrho)$. Then

$$|\log N(mpd) - \log(M + \varrho)| \geq \log \frac{x+1}{x+\frac{1}{2}} \geq \frac{1}{3x}.$$

Therefore, Lemma B.1 shows that

$$\mathbf{1}_{\{N(mpd)\geq M\}} = 1 - \frac{1}{\pi}\int_{-T}^{T}(N(mpd))^{it}\sin(t\log(M+\varrho))\frac{\mathrm{d}t}{t} + O^*\left(\frac{24x}{\pi T}\right).$$

Similarly,

$$\mathbf{1}_{\{N(p)<N(p_s)\}} = \frac{1}{\pi}\int_{-T}^{T}e^{\frac{it}{2}}e^{itN(p)}\sin(tN(p_s))\frac{\mathrm{d}t}{t} + O^*\left(\frac{16}{\pi T}\right),$$

$$\mathbf{1}_{\{N(p)\leq N(p_s)\}} = \frac{1}{\pi}\int_{-T}^{T}e^{-\frac{it}{2}}e^{itN(p)}\sin(tN(p_s))\frac{\mathrm{d}t}{t} + O^*\left(\frac{16}{\pi T}\right).$$

Thus,

$$S_{\overline{\text{II}},2}^{<}(s, \mathscr{X}) = \frac{1}{\pi}\int_{-T}^{T}\sum_{mn\in\mathscr{X}}a_m(t)b_n(t)\frac{\mathrm{d}t}{t} - \qquad (\text{B.16})$$

$$- \frac{1}{\pi^2}\int_{-T}^{T}\int_{-T}^{T}\sum_{mn\in\mathscr{X}}a_m(t,\tau)b_n(t,\tau)\frac{\mathrm{d}\tau\,\mathrm{d}t}{\tau\,t} +$$

$$+ O^*\left(\frac{24x+16}{\pi T} + \frac{16}{\pi^2 T}\int_{-T}^{T}|\sin(\tau\log(M+\varrho))|\frac{\mathrm{d}\tau}{|\tau|}\right) \times$$

$$\times O^*\left(\sum_{\substack{(p_1,\dots,p_s)\in\mathscr{P}_s \\ m:=p_1\cdots p_s}} \sum_{nm\in\mathscr{B}} \sum_{p\in\mathbb{P}_\mathcal{O}(z)} \sum_{\substack{d|\Pi(p) \\ \ell p d=n}} 1\right),$$

with coefficients

$$a_m(t) = \begin{cases} \sin(tN(p_s)) & \text{if } \exists(p_1, \ldots, p_s) \in \mathscr{P}_s : m = p_1 \cdots p_s, \\ 0 & \text{otherwise,} \end{cases}$$

$$b_n(t) = \sum_{p \in \mathbb{P}_{\mathcal{O}}(z)} \sum_{\substack{d \mid \Pi(p) \\ \ell p d = n}} \sum e^{\frac{it}{2}} e^{itN(p)} \mu(d),$$

$$a_m(t, \tau) = a_m(t)(N(m))^{i\tau} \sin(\tau \log(M + \varrho)),$$

$$b_n(t, \tau) = \sum_{p \in \mathbb{P}_{\mathcal{O}}(z)} \sum_{\substack{d \mid \Pi(p) \\ \ell p d = n}} \sum e^{\frac{it}{2}} e^{itN(p)} \mu(d)(N(pd))^{i\tau}.$$

(B.17)

We proceed by gathering some intermediate information before applying (4.9): in the definition of the coefficients b_n, neither of the summations over p and d includes associates. Thus,

$$|b_n(t)|, |b_n(t, \tau)| \leq d((n)).$$

For the other coefficients we always have

$$|a_m(t)|, |a_m(t, \tau)| \leq 1,$$

yet if t and τ are small, one can (and must) do better: indeed, if $|t| \leq x^{-1/2}$ and $|\tau| \leq (\log(x + \frac{1}{2}))^{-1}$, then

$$|a_m(t)| \leq \sqrt{x}|t|, \quad |a_m(t, \tau)| \leq \sqrt{x} \log(x + \tfrac{1}{2})|t\tau|. \tag{B.18}$$

In view of this, we must deal with functions of the type

$$f(t, \delta) = \begin{cases} \delta t & \text{if } |t| \leq \delta^{-1}, \\ 1 & \text{otherwise} \end{cases}$$

and their integrals

$$\int_{-T}^{T} f(t, \delta) \frac{\mathrm{d}t}{|t|} \leq 2\delta \int_0^{\delta^{-1}} \mathrm{d}t + 2 \left| \int_{\delta^{-1}}^{T} \frac{\mathrm{d}t}{t} \right| = 2 + 2 \log(T\delta). \tag{B.19}$$

Lastly, we note that, by Lemma A.9,

$$\sum_{\substack{(p_1, \ldots, p_s) \in \mathscr{P}_s \\ m := p_1 \cdots p_s}} \sum_{nm \in \mathscr{B}} \sum_{p \in \mathbb{P}_{\mathcal{O}}(z)} \sum_{\substack{d \mid \Pi(p) \\ \ell p d = n}} \sum 1$$

$$\leq \#\{\text{units in } \mathcal{O}\} \cdot \sum_{N\mathfrak{a} < x} d_4(\mathfrak{a}) < 6 \cdot 2^{13} x (\log x)^8. \tag{B.20}$$

Collecting what we have gathered so far, we may derive a bound for

$$\mathcal{E} = |S_{\mathrm{II},2}^{<}(s, \mathscr{A}) - \lambda S_{\mathrm{II},2}^{<}(s, \mathscr{B})|$$

as follows: after applying (B.16) with $\mathscr{X} = \mathscr{A}, \mathscr{B}$, the $O^*(\ldots)$-terms are treated directly with (B.20) and (B.19) whereas for the rest one may apply (4.9). Here it is important to use (B.18) for small $|t|$ first, prior to applying (4.9), and (B.19) then bounds the integrals. Therefore, after some computations, we infer

$$
\begin{aligned}
\mathcal{E} \leq\ & Y_{\mathrm{II}} \cdot 2\pi^{-3}(1 + \log(T\sqrt{x})) \times && \text{(B.21)}\\
& \times (\pi^2 + 2(1 + \log(T\log(x + \tfrac{1}{2})))) + \\
& + 3 \cdot 2^{17}\pi^{-1}x(\log x)^8 T^{-1} \times \\
& \times (3x + 2 + 4\pi^{-1}(1 + \log(T\log(x + \tfrac{1}{2})))).
\end{aligned}
$$

Of course, the same arguments also apply to $S^{=}_{\mathrm{II},2}(s, \mathscr{X})$; in view of (B.15) we have to apply this process twice, but in both cases the coefficients corresponding to (B.17) obey the same bounds we used to derive (B.21). Consequently, (B.21) also holds with $S^{=}_{\mathrm{II},2}$ in place of $S^{<}_{\mathrm{II},2}$ if the right hand side is multiplied by two. In total, recalling (B.10), (B.11) and (B.14), we have

$$
|S_{\mathrm{II}}(s)| \leq Y_{\mathrm{II}} + 3 \times \{\text{the bound from (B.21)}\}
$$

and it transpires that choosing $T = x^2(\log x)^8$ already suffices to yield a bound $\ll Y_{\mathrm{II}}(\log x)^2 + 1$. Anyway, to take a little more care of the constants, using a combination of the trivial bounds

$$
x \geq 3, \quad \log(x + \tfrac{1}{2}) \leq \sqrt{x}, \quad (\log x)^8 \leq 4x^{5/2}, \quad x^{-1}\log x \leq \log 3^{1/3},
$$

one easily establishes that $T = 3740x^{9/2}$ yields

$$
|S_{\mathrm{II}}(s)| \leq 800\,(Y_{\mathrm{II}} + 2)(\log x)^2.
$$

On plugging this into (B.9), we thus obtain the assertion of the theorem after recalling (B.5), (B.6) and (B.4).

81

Bibliography

[1] A. G. Abercrombie. Beatty sequences and multiplicative number theory. *Acta Arith.*, 70(3):195–207, 1995.

[2] A. G. Abercrombie, W. D. Banks, and I. E. Shparlinski. Arithmetic functions on Beatty sequences. *Acta Arith.*, 136(1):81–89, 2009.

[3] M. F. Atiyah and I. G. Macdonald. *Introduction to commutative algebra. Student economy edition.* Boulder: Westview Press, 2016.

[4] S. Baier. A note on Diophantine approximation with Gaussian primes, 2016. Preprint: arXiv:1609.08745v3 [math.NT].

[5] A. Baker. Linear forms in the logarithms of algebraic numbers. *Mathematika*, 13:204–216, 1966.

[6] R. C. Baker and L. Zhao. Gaps between primes in Beatty sequences. *Acta Arith.*, 172(3):207–242, 2016.

[7] Th. Bang. On the sequence $[n\alpha]$, $n = 1, 2, \ldots$. Supplementary note to the preceding paper by Th. Skolem. *Mathematica Scandinavica*, pages 69–76, 1957.

[8] W. D. Banks, M. Z. Garaev, D. R. Heath-Brown, and I. E. Shparlinski. Density of non-residues in Burgess-type intervals and applications. *Bull. Lond. Math. Soc.*, 40(1):88–96, 2008.

[9] W. D. Banks and V. Z. Guo. Consecutive primes and Beatty sequences, 2016. Preprint: arXiv:1612.01468 [math.NT].

[10] W. D. Banks and I. E. Shparlinski. Non-residues and primitive roots in Beatty sequences. *Bull. Aust. Math. Soc.*, 73(3):433–443, 2006.

[11] W. D. Banks and I. E. Shparlinski. Short character sums with Beatty sequences. *Math. Res. Lett.*, 13(4):539–547, 2006.

[12] W. D. Banks and I. E. Shparlinski. Prime divisors in Beatty sequences. *J. Number Theory*, 123(2):413–425, 2007.

[13] W. D. Banks and I. E. Shparlinski. Prime numbers with Beatty sequences. *Colloq. Math.*, 115(2):147–157, 2009.

[14] W. D. Banks, A. M. Güloğlu, and C. W. Nevans. Representations of integers as sums of primes from a beatty sequence. *Acta Arith.*, 130(3):255–275, 2007.

[15] W. D. Banks, A. M. Güloğlu, and R. C. Vaughan. Waring's problem for Beatty sequences and a local to global principle. *J. Théor. Nombres Bordeaux*, 26(1): 1–16, 2014.

[16] S. Beatty. Problem 3173. *Amer. Math. Monthly*, 33:159, 1926.

[17] S. Beatty, A. Ostrowski, J. Hyslop, and A. C. Aitken. Solutions to problem 3173. *Amer. Math. Monthly*, 34:159–160, 1927.

[18] A. V. Begunts. On an analogue of the Dirichlet divisor problem. *Vestn. Mosk. Univ., Ser. I*, 2004(6):52–56, 2004.

[19] A. V. Begunts. On the distribution of the values of sums of multiplicative functions on generalized arithmetic progressions. *Chebyshevskiĭ Sb.*, 6(2(14)): 52–74, 2005.

[20] J. Bernoulli III. *Sur une nouvelle espèce de calcul, Recueil pour les astronomes*, volume 1. Berlin, 1771.

[21] A. S. Besicovitch. Sets of fractional dimensions. IV: On rational approximation to real numbers. *J. Lond. Math. Soc.*, 9:126–131, 1934.

[22] S. Bosch. *Algebraic geometry and commutative algebra*. London: Springer, 2013.

[23] J. Brüdern. *Einführung in die analytische Zahlentheorie*. Berlin: Springer, 1995.

[24] J. Brüdern and K. Kawada. The localisation of primes in arithmetic progressions of irrational modulus. *Colloq. Math.*, 123(1):53–61, 2011.

[25] A. S. Fraenkel. The bracket function and complementary sets of integers. *Canad. J. Math.*, 21:6–27, 1969.

[26] A. S. Fraenkel and R. Holzman. Gap problems for integer part and fractional part sequences. *J. Number Theory*, 50(1):66–86, 1995.

[27] J. F. Geelen and R. J. Simpson. A two dimensional Steinhaus theorem. *Australas. J. Combin*, 8:169–197, 1993.

[28] H. Gintner. *Über Kettenbruchentwicklung und über die Approximation von komplexen Zahlen*. PhD thesis, University of Vienna, 1936.

[29] S. W. Graham and G. Kolesnik. *Van der Corput's method of exponential sums*. Cambridge: Cambridge University Press, 1991.

[30] A. M. Güloğlu and C. W. Nevans. Sums of multiplicative functions over a Beatty sequence. *Bull. Aust. Math. Soc.*, 78(2):327–334, 2008.

[31] G. H. Hardy and E. M. Wright. *An introduction to the theory of numbers*. Oxford University Press, 6th edition, 2008.

[32] G. Harman, A. Kumchev, and P. A. Lewis. The distribution of prime ideals of imaginary quadratic fields. *Trans. Am. Math. Soc.*, 356(2):599–620, 2004.

[33] G. Harman. On the distribution of αp modulo one. *J. London Math. Soc. (2)*, 27 (1):9–18, 1983.

[34] G. Harman. On the distribution of αp modulo one. II. *Proc. London Math. Soc. (3)*, 72(2):241–260, 1996.

[35] G. Harman. *Prime-detecting sieves*. Princeton, NJ: Princeton University Press, 2007.

[36] G. Harman. Primes in intersections of Beatty sequences. *J. Integer Seq.*, 18(7), 2015.

[37] G. Harman. Primes in Beatty sequences in short intervals. *Mathematika*, 62(2): 572–586, 2016.

[38] D. R. Heath-Brown. Zero-free regions for Dirichlet L-functions, and the least prime in an arithmetic progression. *Proc. London Math. Soc. (3)*, 64(2):265–338, 1992.

[39] D. R. Heath-Brown. Arithmetic applications of Kloosterman sums. *Nieuw Arch. Wiskd. (5)*, 1(4):380–384, 2000.

[40] D. R. Heath-Brown and C. Jia. The distribution of αp modulo one. *Proc. London Math. Soc. (3)*, 84(1):79–104, 2002.

[41] K. Heegner. Diophantische Analysis und Modulfunktionen. *Math. Z.*, 56: 227–253, 1952.

[42] D. Hensley. *Continued fractions*. Hackensack, NJ: World Scientific, 2006.

[43] A. Hurwitz. Über die Entwickelung complexer Grössen in Kettenbrüche. *Acta Math.*, 11:187–200, 1888.

[44] K. Ireland and M. Rosen. *A classical introduction to modern number theory*. New York: Springer, 2nd edition, 1990.

[45] H. Iwaniec and E. Kowalski. *Analytic number theory*. Providence, RI: American Mathematical Society (AMS), 2004.

[46] V. Jarník. Diophantische Approximationen und Hausdorffsches Maß. *Mat. Sb.*, 36:371–382, 1929.

[47] C. Jia. On the distribution of αp modulo one. *J. Number Theory*, 45(3):241–253, 1993.

[48] C. Jia. On the distribution of αp modulo one. II. *Sci. China Ser. A*, 43(7): 703–721, 2000.

[49] A. Khintchine. Einige Sätze über Kettenbrüche, mit Anwendungen auf die Theorie der Diophantischen Approximationen. *Math. Ann.*, 92:115–125, 1924.

[50] H. D. Kloosterman. On the representation of numbers in the form $ax^2 + by^2 + cz^2 + dt^2$. *Acta Math.*, 49:407–464, 1927.

[51] J. Knopfmacher. *Abstract analytic number theory*. New-York: Dover Publications, 2015.

[52] E. Krätzel. *Lattice points*. Kluwer Academic Publishers, 1988.

[53] L. Kuipers and H. Niederreiter. *Uniform distribution of sequences*. New York etc.: John Wiley & Sons, 1974.

[54] A. V. Kumchev. On sums of primes from Beatty sequences. *Integers*, 8:A8, 12, 2008.

[55] J. Lambek and L. Moser. Inverse and complementary sequences of natural numbers. *Amer. Math. Monthly*, 61(7):454–458, 1954.

[56] E. Landau. Neuer Beweis des Primzahlsatzes und Beweis des Primidealsatzes. *Math. Ann.*, 56:645–670, 1903.

[57] Yu. V. Linnik. On the least prime in an arithmetic progression. I. The basic theorem. *Rec. Math. [Mat. Sbornik] N.S.*, 15(57):139–178, 1944.

[58] Yu. V. Linnik. On the least prime in an arithmetic progression. II. The Deuring–Heilbronn phenomenon. *Rec. Math. [Mat. Sbornik] N.S.*, 15(57):347–368, 1944.

[59] Yu. V. Linnik. New versions and new uses of the dispersion method in binary additive problems. *Sov. Math., Dokl.*, 2:468–471, 1961.

[60] G. Lü and H. Sun. The ternary Goldbach-Vinogradov theorem with almost equal primes from the Beatty sequence. *Ramanujan J.*, 30(2):153–161, 2013.

[61] G. Lü and W. Zhai. The divisor problem for the Beatty sequences. *Acta Math. Sinica (Chin. Ser.)*, 47(6):1213–1216, 2004.

[62] A. A. Markoff. Sur une question de Jean Bernoulli. *Math. Ann.*, 19:27–36, 1881.

[63] K. Matomäki. The distribution of αp modulo one. *Math. Proc. Camb. Philos. Soc.*, 147(2):267–283, 2009.

[64] M. Mkaouar. Beatty sequences and prime numbers with restrictions on strongly q-additive functions. *Period. Math. Hungar.*, 72(2):139–150, 2016.

[65] H. L. Montgomery. *Ten lectures on the interface between analytic number theory and harmonic analysis*. Providence, RI: American Mathematical Society, 1994.

[66] J. Neukirch. *Algebraic number theory*. Berlin: Springer, 1999.

[67] O. Perron. *Die Lehre von den Kettenbrüchen. Band I.* B.G. Teubner Verlagsgesellschaft, 3rd edition, 1954.

[68] O. Ramaré and R. Rumely. Primes in arithmetic progressions. *Math. Comput.*, 65(213):397–425, 1996.

[69] P. Ribenboim. *The new book of prime number records.* New York: Springer, 3rd edition, 1996.

[70] J. B. Rosser and L. Schoenfeld. Approximate formulas for some functions of prime numbers. *Ill. J. Math.*, 6:64–94, 1962.

[71] K. F. Roth. Rational approximations to algebraic numbers. *Mathematika*, 2: 1–20, 1955.

[72] J. Sándor, D. S. Mitrinović, and B. Crstici. *Handbook of number theory. I.* Dordrecht: Springer, 2006.

[73] J. Schmeling, E. Szabó, and W. Reinhard. Hartman and Beatty bisequences. In F. Halter-Koch and R. F. Tichy, editors, *Algebraic Number Theory and Diophantine Analysis: Proceedings of the International Conference held in Graz, Austria, August 30 to September 5, 1998*, pages 405–421. Walter de Gruyter, 2000.

[74] I. E. Shparlinski. Modular hyperbolas. *Jpn. J. Math. (3)*, 7(2):235–294, 2012.

[75] Th. Skolem. On certain distributions of integers in pairs with given differences. *Mathematica Scandinavica*, pages 57–68, 1957.

[76] V. T. Sós. On the distribution mod 1 of the sequence $n\alpha$. *Ann. Univ. Sci. Budap. Rolando Eötvös, Sect. Math.*, 1:127–134, 1958.

[77] H. M. Stark. A complete determination of the complex quadratic fields of class-number one. *Mich. Math. J.*, 14:1–27, 1967.

[78] H. M. Stark. On the 'gap' in a theorem of Heegner. *J. Number Theory*, 1:16–27, 1969.

[79] J. Steuding and M. Technau. The least prime number in a Beatty sequence. *J. Number Theory*, 169:144–159, 2016.

[80] J. W. Strutt. *The Theory of Sound (reprint).* New York: Dover Publications, 1945.

[81] S. Świerczkowski. On successive settings of an arc on the circumference of a circle. *Fundamenta Mathematicae*, 46(2):187–189, 1958.

[82] E. C. Titchmarsh. A divisor problem. *Rend. Circ. Mat. Palermo*, 54:414–429, 1930.

[83] E. C. Titchmarsh. A divisor problem. Correction. *Rend. Circ. Mat. Palermo*, 57: 478–479, 1933.

[84] J. D. Vaaler. Some extremal functions in Fourier analysis. *Bull. Am. Math. Soc., New Ser.*, 12:183–216, 1985.

[85] T. van Ravenstein. The Three Gap Theorem (Steinhaus conjecture). *J. Austral. Math. Soc. Ser. A*, 45(3):360–370, 1988.

[86] R. C. Vaughan. On the distribution of αp modulo 1. *Mathematika*, 24:135–141, 1978.

[87] R. C. Vaughan. The general Goldbach problem with Beatty primes. *Ramanujan J.*, 34(3):347–359, 2014.

[88] I. M. Vinogradov. An elementary proof of a theorem from the theory of prime numbers. *Izvestiya Akad. Nauk SSSR. Ser. Mat.*, 17:3–12, 1953.

[89] I. M. Vinogradov. *The method of trigonometrical sums in the theory of numbers. Translated from the Russian, revised and annotated by K.F. Roth and Anne Davenport. Reprint of the 1954 translation.* New-York: Dover Publications, 2004.

[90] A. Weil. On some exponential sums. *Proc. Nat. Acad. Sci. U. S. A.*, 34:204–207, 1948.

[91] H. Weyl. Über die Gleichverteilung von Zahlen mod. Eins. *Math. Ann.*, 77: 313–352, 1916.

[92] T. Xylouris. *Über die Nullstellen der Dirichletschen L-Funktionen und die kleinste Primzahl in einer arithmetischen Progression*, volume 404 of *Bonner Mathematische Schriften [Bonn Mathematical Publications]*. Universität Bonn, Mathematisches Institut, Bonn, 2011. Dissertation for the degree of Doctor of Mathematics and Natural Sciences at the University of Bonn, Bonn, 2011.

[93] W. Zhai. A note on a result of Abercrombie. *Kexue Tongbao (Chinese)*, 42(8): 804–806, 1997.